Pulsed Laser Processing of Materials

Edited by Dongfang Yang

Published in London, United Kingdom

Pulsed Laser Processing of Materials
http://dx.doi.org/10.5772/intechopen.1002558
Edited by Dongfang Yang

Assistant to the Editor: Katherine Gibson

Contributors
Andra G. Boni, Andreas Hitzler, Colin Veevers, Cristina Chirila, Dana Popescu, Dongfang Yang, Hamed Naderi-Samani, Hao Liu, Hardik Vaghasiya, Jinseng Tang, Kai Han, Katherine Gibson, Lothar Lilge, Lucian D. Filip, Lucian Pintilie, Lucian Trupina, Luminita Hrib, Margarete Akens, Melissa Prickaerts, Mihaela Botea, Minsun Chen, Nick N. Gharabaghi, Paul-Tiberiu Miclea, Raluca Negrea Ioana Pintilie, Reza Shoja Razavi, Robin S. Marjoribanks, Seydi Yavas, Sohret Görkem Karamuk, Thomas Dzelzainis, Viorica Stancu, Weiqiang Yang, Wenxing Xu

First published in London, United Kingdom, 2024 by IntechOpen
IntechOpen is the global imprint of INTECHOPEN LIMITED, registered in England and Wales, registration number: 11086078, 167-169 Great Portland Street, London, W1W 5PF, United Kingdom

British Library Cataloguing-in-Publication Data
A catalogue record for this book is available from the British Library

Additional hard and PDF copies can be obtained from orders@intechopen.com

Pulsed Laser Processing of Materials
Edited by Dongfang Yang
p. cm.
Print ISBN 978-0-85466-611-9
Online ISBN 978-0-85466-610-2
eBook (PDF) ISBN 978-0-85466-612-6

For EU product safety concerns:
IN TECH d.o.o., Prolaz Marije Krucifikse Kozulić 3, 51000 Rijeka, Croatia,
info@intechopen.com or visit our website at intechopen.com.

Meet the editor

Dongfang Yang received his Ph.D. in physical chemistry from the University of Guelph in 1995. He joined National Research Council of Canada in London, Ontario in 2001 and is now a senior research officer. His current research interests include laser materials processing; pulsed laser, sputtering, and e-beam deposition of thin films; new materials development for energy storage devices; chemical and optical sensor development; and electrochemical studies of organic adsorption and self-assembly monolayers. He is an editor or member of the editorial board for over 11 scientific journals and was listed among the top 2% most-cited scientists since 2020, according to a recent Stanford study. He received the IAAM Scientist Award and Medal in 2021 and was named a Vebleo Fellow in 2022.

Contents

Preface

This book covers seven review articles of pulsed laser processing and analyzing of materials. Chapter 1 provides a summary of the use of pulsed lasers in Li-ion battery research, including for the development of large specific surface area nanoparticles of active materials, stable integrative anodes, dense and stoichiometric thin film active materials, as well as electrode architectures facilitating Li-ion diffusion and anode volume expansions and textured current collectors to improve adhesion. The chapter also describes the use of pulsed lasers for cutting and structuring solid ceramic electrolyte and the application of laser ablation inductively coupled plasma mass spectrometry (LA-ICP-MS) for chemical composition analysis of Li-ion batteries throughout their operating cycle. Chapter 2 presents an overview on the fundamentals of ultra-short pulse laser interaction with metals, semiconductors and dielectrics. It reveals the unique responses of the different materials under ultra-short pulse irradiation and various ablation mechanisms due to different photo-energy absorption mechanisms by different materials such as metals, semiconductors, and dielectrics. The two-temperature model is employed to describe the temperature of the electron/carrier and lattice of the materials in nonequilibrium conditions when ultrashort laser pulses are irradiated on them. The chapter also provides an in-depth understanding of the formation mechanisms for the laser-induced periodic surface structures (LIPSS) phenomena. Chapter 3 presents an overview of the properties and applications of epitaxial ferroelectric materials obtained using the pulsed laser deposition (PLD) technique. The chapter reveals that the PLD technique is suitable for obtaining ultrathin films allowing a better understanding of physiochemical phenomena of the ferroelectric materials, materials of significant importance in various electronic and sensing or energy harvesting applications. Chapter 4 presents a review of the use of laser ablation for the synthesis of nanoparticles in liquid for a variety of materials. It discusses in depth the effects of various laser parameters on the size, shape, composition, and crystallinity of nanoparticles. It also reveals the fundamentals, challenges, and limitations of physical and chemical processes involved in PLAL, which facilitates the tailoring of the nucleation and growth of nanoparticles. Chapter 5 presents an overview on the use of ultrashort-pulse burst-mode lasers for processing of materials and biological tissues. Through demonstration of the use of the laser for machining of fused silica, in vitro cell-cultures prepared in hydrogels, and ex vivo articular cartilage, the chapter shows the unique features when this new type of laser is used to process materials as compared to the conventional short duration pulsed laser. Chapter 6 presents an overview of using laser-induced Raman spectroscopy, photoluminescence, and photoluminescence lifetime analysis for gemstone identification. The three laser-based spectroscopy methods allow for differentiation of natural and lab-grown materials, discernment of treatment methods, and determination of the origins of different colored gemstones. Chapter 7 presents an overview on using machine learning methods to help obtain reliable quantitative elemental analysis of materials from highly complicated and nonlinear LIBS spectral data. The chapter introduces a variety of machine learning methods such as PLS,

SVM, LSSVM, Lasso, and artificial neural network-based methods for processing LIBS data and compares the results obtained using those methods in order to provide guidance for the future developments of machine learning methods.

Through these seven chapters written by experts from the international scientific community, the reader will gain a better understanding of the pulsed laser processing and analysis of materials. I gratefully acknowledge all chapter authors for their enthusiastic and collaborative contributions and would like to thank Ms. Iva Horvat and Mr. Tonci Lucic, publishing process managers, for their guidance and support in the preparation of this book.

Dongfang Yang, Ph.D.
Senior Research Officer,
Automotive and Surface Transportation,
National Research Council Canada,
London, Ontario, Canada

Katherine Gibson, M.E.Sc.
Research Associate,
Automotive and Surface Transportation,
National Research Council Canada,
London, Ontario, Canada

Section 1

Laser Processing of Materials

Chapter 1

Applications of Pulsed Laser Ablation in Li-Ion Battery Research

Katherine Gibson and Dongfang Yang

Abstract

Harnessing pulsed laser ablation processes in the manufacturing of energy storage devices is a new and promising strategy for the facile development of next-generation Li-ion batteries. In laser ablation, a pulsed laser is focused on a material surface such that the transfer of energy causes the removal of localized material via high throughput and environmentally-friendly processing. This chapter will provide a summary of the recent advances in laser ablation technologies for producing Li-ion battery materials and components. In terms of electrode optimization, it will examine the use of pulsed lasers to: (1) generate large specific surface area nanoparticles of active materials or stable integrative anodes; (2) deposit compositionally complex and stoichiometric thin film active materials; (3) create electrode architectures with increased Li-ion diffusion kinetics, enhanced wettability or free space to accommodate Si anode volume expansions, and; (4) remove the superficial inactive or solid electrolyte interface layers from electrode surfaces. It will also investigate the laser ablation of current collectors to produce textures with improved adhesion and the use of pulsed lasers for cutting and structuring solid ceramic electrolyte. Finally, this chapter will discuss the application of laser ablation inductively coupled plasma mass spectrometry (LA-ICP-MS) for chemical composition analysis of Li-ion batteries throughout their operating cycle.

Keywords: pulsed lasers, Li-ion batteries, laser ablation, energy storage devices, pulsed laser deposition, laser micromachining, LA-ICP-MS

1. Introduction

Mounting concerns associated with global warming and environmental degradation have resulted in increased efforts towards electrification. In particular, ground transportation is responsible for 15% of worldwide CO_2 emissions [1] and therefore constitutes one of the most consequential candidates for electrification. The development of energy storage technologies is vital in accommodating the transition towards electrical propulsion. Due principally to their high specific energy densities, long cycle lives and high Coulombic efficiencies, lithium-ion (Li-ion) batteries are currently considered the most suitable energy storage candidates for electric vehicle (EV) applications [2]. As a result, significant resources have recently been placed over the last two decades on optimizing Li-ion battery technologies.

1.1 Li-ion batteries

Li-ion batteries rely on the concept of extremely low electrode potential and the tendency of a Li atom to lose its single valence electron to become a charged Li^+ ion. **Figure 1** provides a schematic of the Li-ion battery operating mechanism. A typical Li-ion battery consists of an Al-foil current collector connected to a Li transition-metal oxide cathode and a Cu-foil current collector connected to a graphite anode. The cathode is a supplier of Li^+ ions, while the anode is a storage medium for Li^+ ions. Electrolyte positioned between the cathode and anode allows for the flow of Li^+ ions while inhibiting the passage of electrons. To prevent short-circuiting, an insulating separator is included, with pores for Li^+ permeation. In the charge state, a Li-ion battery is charged by applying a voltage by a power source which attracts the metal oxide valence electrons to its positive terminal. These electrons flow from the metal oxide cathode, through the external circuit to reach the graphite anode of the battery. Simultaneously, the Li^+ ions are attracted to the negative terminal of the power supply and flow through the electrolyte to reach the anode. A Li-ion battery is considered to be fully charged once all of the active surface sites of the anode are occupied by the Li^+ ions. However, this is an unstable state. As a result, the battery can act as an energy source (i.e., a galvanic device) by applying a load which allows the electrons to flow once again through an external circuit to return to the cathode. To maintain neutrality, the Li^+ ions are released from the anode and flow through the electrolyte and separator to the cathode. Li^+ intercalation facilitates charging and discharging of Li-ion batteries. In the charge state, the cathode is oxidized and delithiated while the anode is lithiated. Similarly, in the discharge state, the anode is delithiated while the cathode is reduced and lithiated.

Li-ion battery research over the last two decades has looked to optimize all battery components through materials selection and modification in order to achieve improved performance, cost and operational safety. Since energy density is limited by the cathode (when conventional graphite anodes are used), the optimization of cathode materials has received significant attention. In particular, Co-, Mn-, and Ni-based materials have been investigated for their high electrochemical potentials which permit large battery cell voltages [2]. Polyanionic PO_4^- based cathodes have also been explored since they are significantly less susceptible to thermal runaway [3]. Even organic photoactive cathodes with improved specific capacity under illumination have

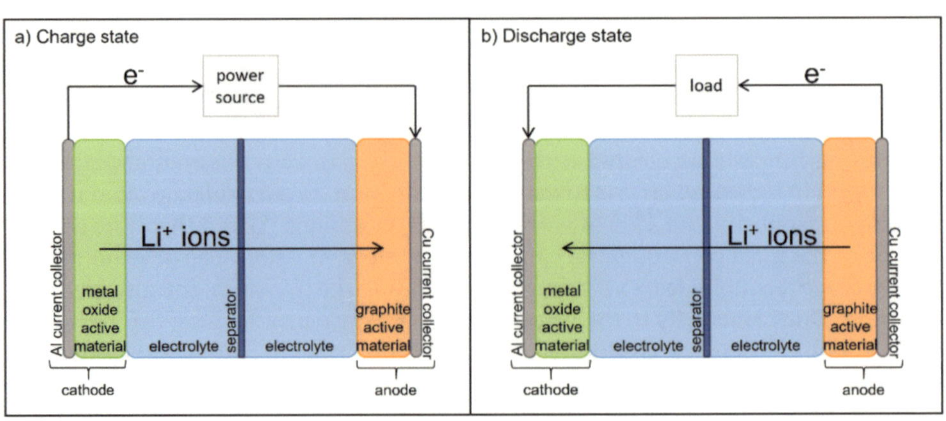

Figure 1.
Schematic of a Li-ion battery operating in the (a) charge and (b) discharge states.

been reported [4]. Evidently, cathode optimization research is ongoing as researchers look to further improve ionic and electronic conductivity, inhibit the formation of passivation layers, reduce costs and support operational safety [5]. In terms of anode optimization, Si- and Sn-materials with higher specific capacities than graphite have been proposed. In particular, Si-based anodes enable the highest theoretical capacity—4200 mAh/g—approximately 10 times higher than that of graphitic carbon anodes. However, severe volume expansion under lithiation limits their practical application due to structural compromise which results in poor electrical contact with the current collector [5]. Li-metal has similarly been investigated as an anode material for its high theoretical capacity and low electrochemical potential, however, dendrite formation causing short circuiting threatens device safety [2]. Solid-state electrolyte must therefore be employed with Li-metal anodes, but low ionic conductivity, as well as rigidity and roughness yield compromised contact resistance. **Table 1** summarizes the key advantages and disadvantages of various Li-ion battery anode active materials. In terms of current collectors, efforts are centered around reducing foil thickness in order to reduce the total weight of devices. However, thin current collectors sacrifice electrical conductivity and suffer from poor mechanical integrity, which yields increased contact resistance with cycling [6]. Therefore, it is clear that while Li-ion batteries have already been deployed commercially, they stand nevertheless to gain from further research and development.

1.2 Pulsed laser ablation

Pulsed laser ablation is a commercially-relevant method for synthesizing and modifying materials and components in next-generation Li-ion batteries. In pulsed laser ablation, a laser beam is focused on a material or component surface. Absorption of the laser energy results in the sublimation and removal of superficial material. Ultrafast pulse durations limit heat conduction, leaving the surrounding material in pristine condition, without physical or chemical alteration. At high flux, the laser will excite and ionize the plume of ejected material, forming plasma and inciting optical emissions [7].

Laser ablation can be harnessed in two modes, where the desired product is either: (1) the ejected material, or; (2) the ablated solid surface. Ejected material is the desired product in the generation of nanoparticles and thin films by pulsed laser ablation. In each case, the laser parameters are tuned to eject material with specific morphologies. Conversely, machining is the most common application of pulsed laser

Anode active material	Key advantages	Key disadvantages
Graphite	Commercial Li-ion battery anode active material due to very long cycle life and good energy density	Energy density is not as high as alternative materials
Si	Very high specific capacity	Poor electronic conductivity and volume expansion under lithiation limits cycle life
Sn	Very high specific capacity	Volume expansion under lithiation limits cycle life
Li metal	Very high specific capacity and low reduction potential	Solid state electrolyte required to prevent short-circuiting

Table 1.
Comparison of the key advantages and disadvantages of various Li-ion battery anode active materials.

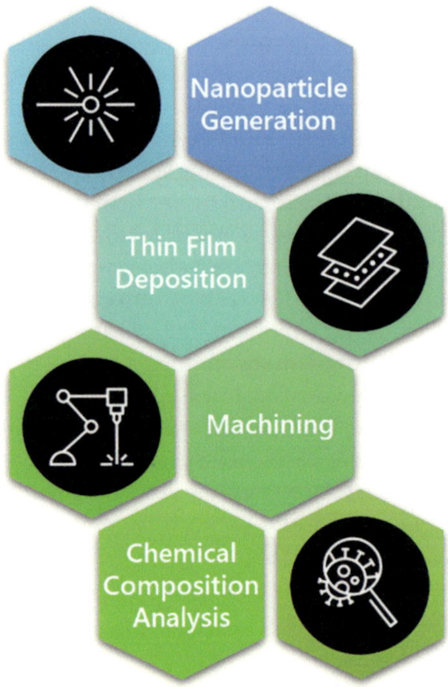

Figure 2.
Outline of pulsed laser application areas in Li-ion battery research.

ablation where the desired product is the ablated solid surface. Importantly, laser ablation permits high throughput and environmentally-friendly processing of materials and can easily be tuned by manipulating the pulse duration, pulse frequency, wavelength, energy density and scanning speed of the laser.

In Li-ion batteries, pulsed laser ablation has been employed for synthesis, modification and analysis of materials and components via: (1) nanoparticle generation; (2) thin film deposition; (3) machining, and; (4) chemical composition analysis. As depicted in **Figure 2**, this chapter will examine the applications of pulsed laser ablation in next-generation Li-ion battery research.

2. Pulsed lasers for nanoparticle generation

Pulsed lasers provide a robust and facile means of generating nanoparticles to be used as large surface area active materials in Li-ion battery anodes and cathodes. In general, the push towards smaller-sized materials comes as a result of the Li^+ insertion/extraction processes at the anode and cathode, which are limited by the rate of Li^+ diffusion. By incorporating nano-sized active materials at the cathode and anode, shorter lengths are obtained for both electronic and ionic transport. Compared to the conventional methods, in the laser-based process, no chemical additives or reagents are required (although surfactants can be added to prevent agglomeration), and there are no reaction by-products. Nanoparticles can either be generated by a two-step pulsed laser ablation in liquid (PLAL) process or a single-step in-situ laser ablation oxidation process.

2.1 Pulsed laser ablation in liquid for nanoparticulate active materials

In PLAL, the laser beam is focused on a solid target material submerged in liquid media and the ablated material is dispersed into the liquid [8]. **Figure 3** provides a schematic of the PLAL process. Importantly, PLAL permits for recovery of the ejected material in nanoparticulate powder form. In other words, PLAL is distinguished in that the generated nanoparticles remain detached from either the target material or a substrate due to their suspension in liquid media. In order to recover the nanoparticles, a second separation step, involving drying or centrifugation, is required.

In 2007, Tsuji et al. employed PLAL to generate $LiMn_2O_4$ nanoparticulate active material for improved cathode performance [9]. In particular, they hypothesized that the discharge capacity would improve in the large surface area nanoparticulate active material samples as a result of a decrease in Li diffusion distance. Tsuji et al. irradiated a suspension of $LiMn_2O_4$ powder in water with a Nd:YAG laser at 150 mJ/pulse for 1 hour and observed a 90% reduction in particle size from 5 μm to <100 nm. However, they did not observe any improvements in discharge capacity for current densities above 10 mA/cm^2 and reported a 25% reduction in discharge capacity for current densities below 10 mA/cm^2. The unexpected result is explained by the degradation of $LiMn_2O_4$ to MnO and Li, and Li/Mn cation mixing under laser ablation. Nonetheless, this work is significant in that it constitutes one of the first of its kind to employ PLAL for nanoparticulate active materials in Li-ion batteries and serves as a starting point to demonstrate the potential and opportunities for improvement of this method.

In 2015, Nowak et al. utilized PLAL to optimize Li-ion battery anodes via doping with SnO nanoparticles to increase the specific capacity [10]. The SnO was synthesized by focusing a Nd:YAG laser for 45 minutes on a metallic Sn target submerged in water. An energy density of 20 J/cm^2, a repetition rate of 10 Hz, and a pulse duration of 6 ns were used as ablation parameters. Subsequently, the SnO nanoparticles were embedded in a carbon matrix by mixing with gelatine and pyrolyzing at 900°C to form the $SnO@C_{gel}$ anodic active material. The final electrode incorporated 60% active material, 20% carbon black and 20% binder, by mass. Importantly, the selected active material demonstrated improved performance as a result of: (1) the PLAL-generated SnO nanoparticles, which enhanced capacity, and; (2) the organic matrix, which provided stable cyclability. After 140 cycles at a current density of 100 mA/g,

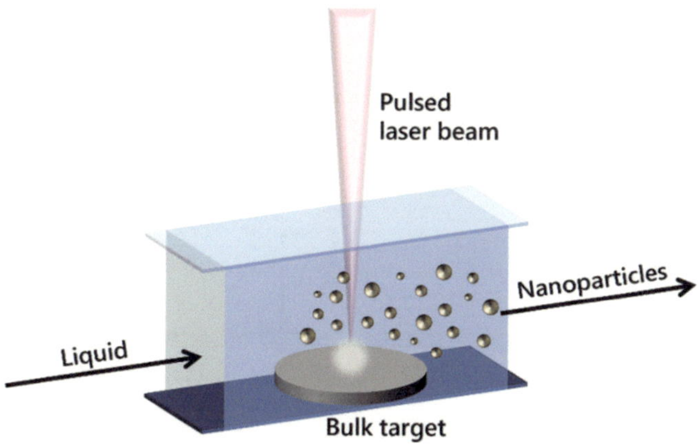

Figure 3.
Schematic of the pulsed laser ablation in liquid (PLAL) process.

Nowak et al. reported a capacity of 580 mA h/g for their Li-ion battery, corresponding to a 98% retention in capacity.

Given the importance of Li-ion batteries in supporting transportation electrification, it has become crucial to devise methods of extending cycle life. The formation of a solid electrolyte interface (SEI) passivation layer on the anode, which restricts Li^+ insertion, is to blame for compromised long-term capacity stability. To overcome this issue, $Li_4Ti_5O_{12}$ (LTO) has been investigated as an alternative to conventional graphite anodes for its high working potential which can suppress SEI layer formation. However, LTO suffers from small Li^+ diffusion coefficients, limiting capacity. Recently, Alrefaee et al. investigated the use of PLAL to reduce LTO particle size and improve Li^+ diffusion [11]. They submerged a solid LTO target in liquid polyethersulfone (PES) and ablated it with a Nd:YAG laser at a repetition rate of 10 Hz and a pulse duration of 7 ns for 60 min. As shown in **Figure 4(a)–(c)**, by changing the energy from 40 mJ/pulse to 120 mJ/pulse, Alrefaee et al. were able to reduce the size of LTO nanoparticles from 26 nm to 5 nm. Electrochemical testing demonstrated superior

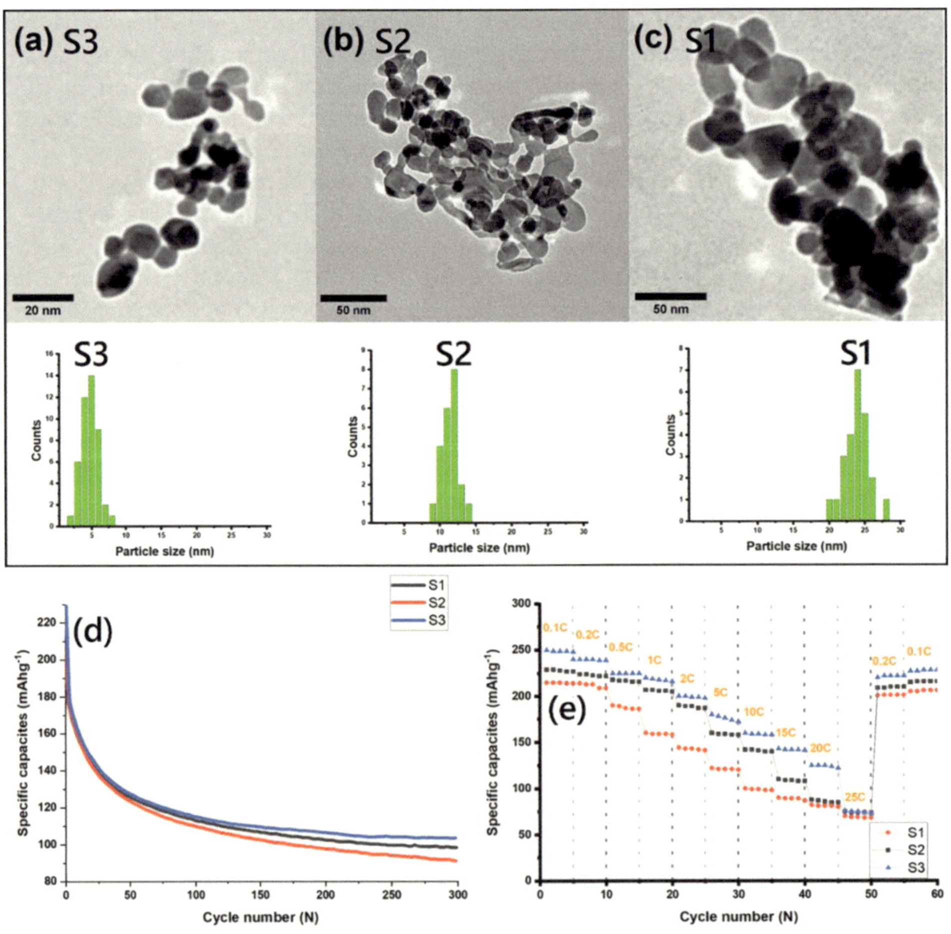

Figure 4.
TEM images of LTO nanoparticles generated with laser energy of: (a) 40, (b) 80, and (c) 120 mJ/pulse, (d) cycle performance at 1C and (e) rate capability of electrochemical cells with different with LTO particle sizes of 26 nm (S1), 12 nm (S2) and 5 nm (S3). Reprinted with permission from [11]. Copyright (2023) Elsevier.

performance with reduced particle size: in **Figure 4(d)**, initial discharge capacities at 1 C were 180 and 244 mA h/g for LTO particle sizes of 26 and 5 nm, respectively. Similarly, in **Figure 4(e)**, after cycling at high rates, discharge capacities remained the highest for the smallest LTO particle size.

2.2 Laser ablation oxidation for metal oxide integrated anodes

In laser ablation oxidation, a nanoparticulate metal-oxide layer, XO_x, is grown on its pure metal precursor, X, to from an integrated XO_x-X material. As depicted in **Figure 5**, growth arises when: (1) the pulsed laser ablates a target metal surface, vapourizing and/or melting atoms under its high energy-intensity radiation; (2) the vapor/liquid-phase metallic atoms collide with oxygen atoms under atmospheric conditions, resulting in a loss of kinetic energy, atomic clustering, solidification and oxidation, and (3) the solid metal-oxide clusters are re-deposited onto the target metal surface. Laser parameters such as the frequency, scanning speed, pulse energy and pulse width can be tuned to control the size of metal oxide nanoparticles. Importantly, the use of lasers for integrated nanoparticulate metal-oxide growth is economical due to its simplicity and scalability. Additionally, the growth mechanism, which relies predominantly on the photothermal transfer of laser radiation to vapourize a metal target, is robust and widely applicable across metallic materials.

The principles of laser ablation oxidation were first applied in Li-ion battery research in 2015 by Zhong's group from Tsinghua University, as a means of accommodating the volume expansion plaguing stability in high-capacity Si-based anodes. The porous structures in silicon-based nanomaterials provide empty space for volume expansion under lithiation, reducing strain and improving cyclability. However, the conventional wet chemical- or chemical vapor deposition-based synthesis techniques do not permit the simultaneous achievement of high-throughput conditions, while ensuring precise microstructural control. In response, Zhong et al. introduced the laser ablation oxidation of monocrystalline Si wafers to form SiO_x-Si integrated anodic active material for Li-ion batteries, using a pulsed laser (λ = 532 nm) with a frequency of 1000 Hz, a pulse width of 10 ns and an energy of 1.5E-4 J [12]. Using this novel method, their porous SiO_x-Si anodes achieved a high initial discharge capacity of 1400 mAh/g, and retained a capacity of 960 mAh/g after 800 cycles (more than double the capacity attained after 800 cycles with commercial graphite anodes). Similar work has since been replicated using an IR laser [13].

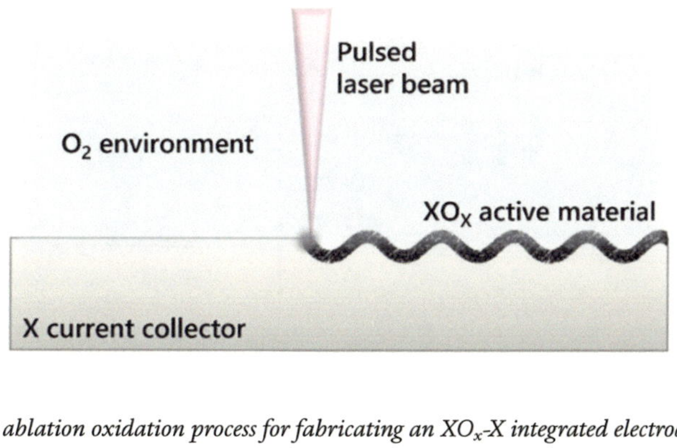

Figure 5.
Schematic of the laser ablation oxidation process for fabricating an XO_x-X integrated electrode.

Building on their previous work, Zhong et al. discovered a new application for laser ablation oxidation in Li-ion batteries: the *in-situ* growth of metal oxide active materials on metallic current collectors [14]. This evolution of the laser ablation technology takes advantage of the existing porous metal oxide nanostructures to prevent strain upon volume expansion, but also renders unnecessary the use of binder, thereby improving the adhesion of active material to the current collector and the electronic conduction of the anode. They used the same 532 nm laser with a frequency of 30 kHz, a pulse width of 12 ns and a pulse energy of 0.4 mJ to ablate Cu foil, forming a CuO-Cu integrated anode. Remarkable stability was achieved: the capacity after 800 cycles at 1.5 A/g was 394 mAh/g, corresponding to a Coulombic efficiency greater than 99%. In subsequent publications, Zhong et al. demonstrated the robustness of the laser ablation oxidation technique, reporting the preparation of binder-free CoO-Co, NiO@C-Ni, Fe_2O_3/Fe_3O_4@C-Fe, and MoO_3-Mo integrated anodes [15, 16].

3. Pulsed lasers for thin film deposition

Pulsed lasers can be used to grow 2D materials in a widely employed and mature technique known as pulsed laser deposition (PLD). As shown in **Figure 6**, PLD is a vacuum-based deposition method, wherein a pulsed laser beam is focused on the surface of a solid target material. The laser photons are converted into thermal, chemical and mechanical energy, resulting in the rapid extraction of material from the surface of the target to form a plasma plume which condenses on the substrate [17].

The PLD technique can be used to produce high-quality and dense films with tunable morphologies. However, since large-scale deposition is challenging, PLD may be particularly well-suited to the manufacturing of thin film batteries. Given the prevalence of microelectronics in the technology market, there is a heightened need for small, lightweight batteries, offering long lifetimes and high energy densities. Solid-state thin film batteries are a proposed solution, wherein all battery components are thin films that amount to a multilayer device. Physical vapor deposition techniques such as PLD are attractive in the fabrication of thin film batteries because they negate the need for binder, thereby improving energy density.

The PLD technique yields highly uniform films with preserved stoichiometry; it is ideally suited for producing novel thin film compositions and structures. The fabrication of compositionally complex thin films containing lithium is particularly

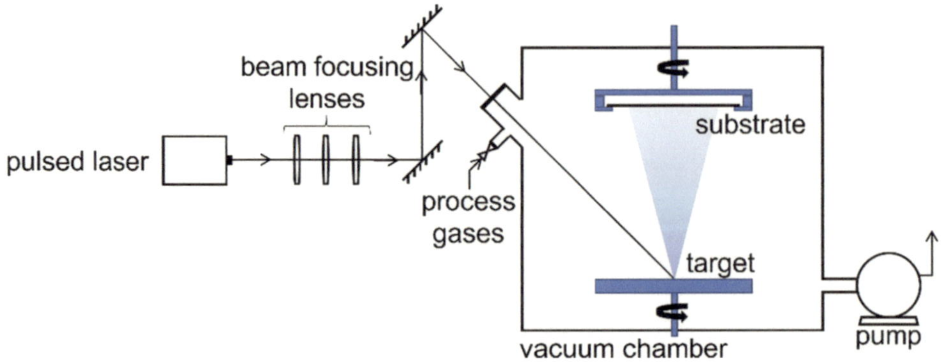

Figure 6.
Schematic of a pulsed laser deposition system.

challenging because lithium is a very easily evaporated element, rendering stoichiometric preservation difficult by other deposition methods. As a result, PLD is considered one of the best techniques for producing thin film cathode active materials.

$LiCoO_2$ is considered one of the best cathode active materials because of its high energy density and superior cycling stability. As such it is already being employed in commercial Li-ion secondary batteries. Antaya et al. published one of the first instances of using the PLD technique under optimized processing conditions to produce nominally stoichiometric $LiCoO_2$ thin films with near-zero Li loss [18]. They employed a pulsed XeCl laser (λ = 308 nm) with a fluence of 5 to 6 J/cm^2 to ablate a solid $LiCoO_2$ target under oxygen atmosphere. A post-deposition annealing step at temperatures up to 700°C was applied to produce crystalline films. The authors studied the effects of crystallinity and on the electrochemical properties and found that the thin films of $LiCoO_2$ annealed above 500°C had the largest fractions of high-temperature $LiCoO_2$ crystalline phases and correspondingly exhibited the highest capacities.

Since PLD permits the facile deposition of compositionally complex thin films, researchers have used it to further tune the chemical composition of conventional $LiCoO_2$ active materials. In order to increase battery voltage and reduce toxicity, Perkins et al. substituted half of the Co in $LiCoO_2$ for Al, producing thin films of $LiCo_{0.5}Al_{0.5}O_2$ [19]. Al substitutions are also economical in that Al is lighter and less expensive than Co. They employed the PLD technique to ablate stoichiometric ceramic targets of $LiCo_{0.5}Al_{0.5}O_2$ using an excimer laser (λ = 248 nm) at 10 Hz and 325–415 mJ/pulse. The authors found that $LiCo_{0.5}Al_{0.5}O_2$ cathodes exhibit higher voltages during charging cycles than $LiCoO_2$ cathodes, however they report significant hysteresis between charging and discharging cycles, ultimately resulting in a 200% reduction in capacity for the $LiCo_{0.5}Al_{0.5}O_2$ cathodes.

While PLD typically yields thin films that are stoichiometrically equivalent to the target material, due to the volatility of the atoms involved, Li and O deficiencies are not uncommon. This is particularly true in the deposition of $LiMn_2O_4$, which is of interest in micro-batteries for its safer overcharge tolerance. In order to overcome compositional deficiencies, the PLD parameters such as the deposition temperature, oxygen pressure and substrate-to-target distance should be tuned. For instance, in order to prevent Li deficiencies that occur as a result of Li-out diffusion, films should be grown at substrate temperatures less than 400–600°C [20, 21]. Moreover, Li deficiency is increased when the distance between the substrate and the target is large [22]. Interactions between laser-produced plasma and oxygen background gas are not well-understood. However, it is generally agreed upon that Li and O deficiencies emerge in films grown under high vacuum or low (<10 Pa) oxygen pressures [20, 23]. This may be related to a slowing of the laser-produced plasma ions with increasing background gas pressures and a preferential re-sputtering of Li under vacuum or low background gas conditions [20].

PLD has also been used to optimize Li-ion battery cathodes via the addition of functional thin film surface layers. As a result of electron exchange during operation, a solid electrolyte interface (SEI) layer forms at the $LiCoO_2$ cathode-electrolyte interface, increasing the resistance to further charge transfer and limiting capacity retention. Teranishi et al. used the PLD technique to cover their $LiCoO_2$ active material with $BaTiO_3$, thereby suppressing the formation of native SEI [24]. They were able to achieve three-dimensional $LiCoO_2$ coverage by modifying the conventional PLD configuration: as shown in **Figure 7(b)**, the plasma plume was directed into a dynamically-mixed pan containing $LiCoO_2$ powder, such that $BaTiO_3$ nanoparticles were deposited onto the $LiCoO_2$ particle surfaces. The dielectric $BaTiO_3$ was

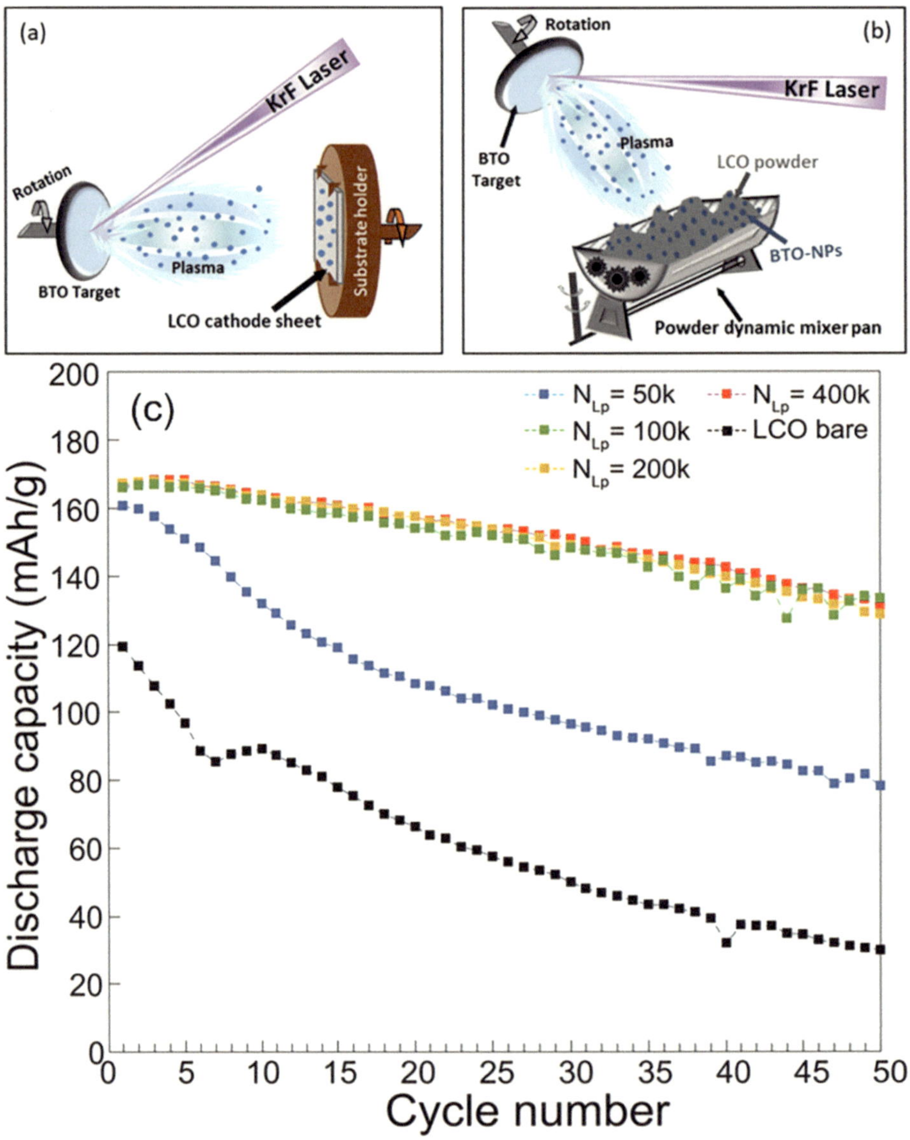

Figure 7.
Schematic representation of (a) the conventional PLD method, and (b) the PLD-based nanodecoration of $BaTiO_3$ on LCO powder, (c) Cycle performance at 10 C of $BaTiO_3$ nanodecorated LCO after different numbers of laser ablation pulses (N_{Lp}) compared to bare LCO. Reprinted with permission from [24]. Copyright (2022) AIP Publishing.

multifunctional in that it not only prevented SEI formation, but also provided pathways with minimized activation energy for Li migration. Electrochemical performance testing, given in **Figure 7(c)**, demonstrated both improved rate capability and capacity retention with $BaTiO_3$ incorporation by PLD. Notably, after cycling at a high current rate of 10 C for 50 cycles, the PLD $BaTiO_3$ cathodes retained ~ 80% capacity, compared to only 25% for those without $BaTiO_3$ decoration.

While the compositionally complex nature of lithium transition metal oxide cathode active materials makes them ideal candidates for fabrication via the PLD technique, all thin film battery components can be grown by PLD. As a result, the PLD of anode materials [25–28] and solid electrolyte [29–31] has also been investigated extensively. In fact, to achieve the most economical thin film batteries, all battery layers should be deposited by a single technique, without breaking the vacuum. In this pursuit, Kuwata et al. were the first to report the fabrication of a complete thin film Li-ion battery using only the PLD technique [32]. They employed a Nd:YAG laser (λ = 266 nm) at a frequency of 10 Hz and a fluence of 3.5 J/cm^2 to sequentially ablate various target materials under oxygenated environment. Following this methodology, the authors successfully demonstrated the production of an all-PLD thin film battery consisting of a: (1) Pt/Cr cathode current collector; (2) $LiCoO_2$ cathode active material; (3) Li_2O-V_2O_5-SiO_2 (LVSO) solid electrolyte; (4) SnO anode active material, and (5) Pt anode current collector.

4. Pulsed lasers fvor machining

Instead of adding functional material layers or nanoparticles, pulsed lasers can equally be used in the removal of material to create textures or microstructures. Also known as laser micro-milling or laser micromachining, the use of pulsed lasers to accurately machine fine features and complex geometries into a material surface is important in industries and applications requiring a degree of precision that is not achievable by conventional mechanical machining [33]. In laser micromachining, a pulsed laser with specific frequency, fluence and beam radius is scanned across a material surface. Ultra-fast pulse frequencies ensure that photons are absorbed faster than heat can be diffused out of the irradiated region, resulting in localized ablation without any damage to the surrounding area. For this reason, low-melting temperature materials such as plastics can be laser micromachined. Fluence can be increased to achieve higher rates of ablation and deeper textures, however excess fluence can compromise quality by heating the material surface. Finally, the radius of the laser beam will directly determine the machining resolution; by manipulating the focusing optics, machining resolutions from hundreds of nanometers to hundreds of microns can be achieved. The type of laser must also be carefully selecting according to the application. In particular, the use of UV lasers is common because UV light is readily absorbed by most materials. Importantly, laser micromachining is positioned as a green manufacturing process: since the heat-affected zone produced during laser micromachining is limited, energy is used efficiently and consumption is minimized, the need for cooling water is eliminated, and post-processing steps such as polishing are rendered unnecessary.

4.1 Creation of 3D architectures

The microstructuring of electrode materials to increase electrode surface area and enhance Li-ion diffusion kinetics is considered an important avenue for next-generation Li-ion battery development. An increasingly electric society has called for batteries with improved energy density. The use of thick electrodes with an increased ratio of active material to inactive current collector and separator foils has been proposed as a straightforward solution. However, there exists an inverse relationship between specific energy and specific power, such that thick electrode batteries often exhibit increased resistance to diffusion, compromising rate capability and long term cyclability. In order to overcome the challenges associated with thick electrodes, pulsed laser micromachining has been employed to generate 3D active material architectures with enhanced diffusion kinetics. A schematic of the electrode laser-structuring process is given in **Figure 8(a)**.

This has been demonstrated for many cathode active materials, including $LiMn_2O_4$ [35, 36], $LiFePO_4$ [37] and $Li(NiMnCo)O_2$ (NMC) [34, 38, 39]. For instance, Pröll et al. used a fiber laser at λ = 515 nm with a pulse energy and width of 0.125 μJ and 350 fs, respectively, to create 3D grid architectures with an aspect ratio of ~ 1.7 in 60 μm thick $LiMn_2O_4$ cathode active material [35]. They also employed calendering to improve electrical contact with the current collector. Galvanostatic and cyclic voltammetric testing confirmed the improved diffusion kinetics of the laser-structured electrodes, ultimately resulting in a 50% increase in capacity retention at 1C compared to its untreated counterpart.

Zhu et al. employed a similar approach for simultaneous optimization of energy and power density in Li-ion batteries with $Li(Ni_{0.6}Mn_{0.2}Co_{0.2})O_2$ (NMC 622) cathodes [34]. Line-patterning NMC622 cathodes with a λ = 1030 nm laser (SEM image given in **Figure 8(b)**) revealed the facile tuning of battery performance for various applications: (1) thick (>200 μm), non-laser-textured cathodes provide the largest energy density; (2) thin (<100 μm) laser-structured cathodes provide the highest power, and; (3) thick, laser-structured cathodes provide both high power and high energy density. Ragone plots demonstrating these findings are given in **Figure 8(c)**.

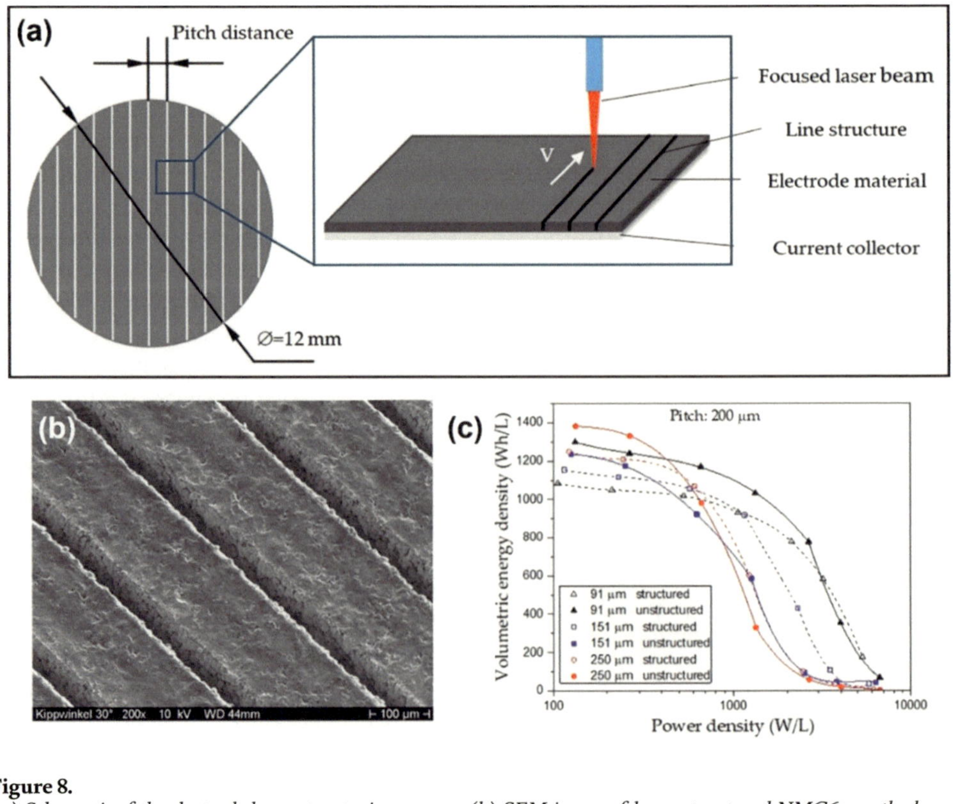

Figure 8.
(a) Schematic of the electrode laser-structuring process. (b) SEM image of laser-structured NMC622 cathodes (91 μm of active material on Al current collector) with a 200 μm pitch distance. (c) Ragone plots for unstructured and laser-structured cells with active material thicknesses of 91, 151 and 250 μm. Reproduced with permission from [34]. Copyright (2019) MDPI.

Laser surface modification technologies have also garnered significant attention across biomedical, oil, automotive, etc. industries for their ability to change surface wetting characteristics. Compared to conventional chemical-based techniques, tailoring wettability via laser surface modification is more environmentally friendly and long-lasting [40]. Surface wettability is of particular interest in Li-ion batteries because Li-ion diffusion only occurs across wetted electrode surfaces. Moreover, electrolyte filling due to poor surface wettability is a major bottleneck in the manufacturing of batteries. As a result, laser surface modification technologies have been harnessed to increase electrode wettability. Kleefoot et al. used a low-energy nanosecond pulsed laser to selectively thermally decompose the amorphous binder and carbon black phases in graphite active materials, thereby increasing the surface concentration of crystalline graphite particles [41]. The result is a roughened surface without a loss of active material. Importantly, by roughening the surface of graphite active materials via the aforementioned methodology, Kleefoot et al. reported a pronounced 8-fold increase in wetted surface area.

Finally, pulsed lasers can be used to create electrode architectures with free space or porosity to accommodate Si anode volume expansions, thereby advancing silicon-based anode implementation. Zheng et al. demonstrated that mechanical stress in silicon/graphite composite electrodes is significantly reduced as a result of laser-generated electrode structuring, leading to improved capacity retention, cycle stability and cell lifetimes [42].

While significant improvements in battery performance have been demonstrated with laser-texturing of electrodes, the question of commercial viability remains. In particular, in order to avoid bottlenecks in manufacturing, laser patterning speeds of several m/min must be demonstrated. With the objective of scaling the laser ablation technologies for commercial battery manufacturing, Meyer et al. studied the impact of laser processing parameters on resulting graphite and silicon/graphite electrode architectures [43]. As demonstrated in **Figure 9**, they observed an important trade-off between laser scanning speed and aspect ratio. Nonetheless, their investigation revealed that laser power can be increased to compensate for the reduction in aspect ratio. Notably, Meyer et al. were able to demonstrate high aspect ratio (~ 1.8) electrode texturing through the active material layer and down to the current collector at laser powers of 180 W for fast scanning speeds up to 1.7 m/s (as indicated by the red arrow in **Figure 9**).

In an effort to improve data accessibility and accelerate commercialization, Tancin et al. have recently published a thorough comparative study of the effect of laser ablation parameters on the resulting ablated microstructures for several electrode materials including graphite, silicon, NMC and LTO [44]. Importantly, the authors examined industrially-relevant ablation metrics such as repeatability, ablation rate and aspect ratio.

4.2 Regeneration via surface layer removal

In place of partial surface layer removal—as in the creation of 3D architectures—pulsed laser ablation can be similarly employed to remove the entire surface layer of a component. This form of ablation is appropriate when components are contaminated during operation or rendered inactive due to chemical reactions with their operating environment. In fact, lasers are commonly used in manufacturing industries to extend the service life of parts by removing mechanically or electrically compromised surfaces before failure occurs. Often termed "laser cleaning," this process is common

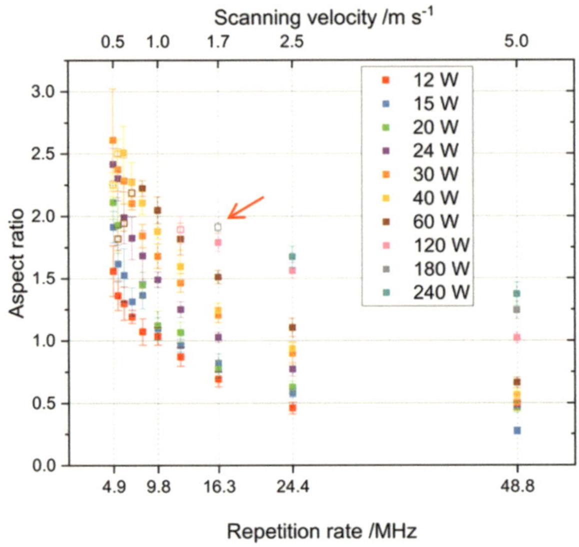

Figure 9.
Aspect ratio of the grooves in laser-microstructured silicon/graphite composite electrodes as a function of laser scanning velocity or repetition rate for laser powers from 12 to 240 W. Solid symbols indicate that the active material was textured through to the current collector, while open symbols indicate that texturing did not reach the current collector. Reproduced with permission from [43]. Copyright (2023) Laser Institute of America.

across semiconductor, welding and shipbuilding industries to remove oxide layers such as rust and contaminants such as millscale, salt and oil [45].

Currently, commercial Li-ion batteries are plagued by capacities that decrease over their lifetimes. For instance, in EV batteries, capacities degrade by an average of 2.3% annually [46]. Since these batteries are considered to have reached end-of-life by the time they degrade to 70–80% of their original capacity [47], lifetimes of less than 9 years are to be expected. This degradation is predominantly a result of the formation of solid electrolyte interface (SEI) layers on battery anodes, which increase resistance and trap lithium ions, reducing the total amount of lithium available for intercalation. The SEI is composed of oxidation and reduction reaction products, namely Li_2CO_3 [48] and LiF [49], depending on the electrolyte used. Chemical methods have been shown to effectively remove SEI and restore capacity [50], however their toxicity limits adoption. Pulsed laser ablation therefore emerges as a sustainable alternative for Li-ion battery regeneration. Moreover, the small heat affected zones generated under ablation with ultrafast pulsed lasers prevents energy dissipation and damage to subsurface layers.

Zhang et al. demonstrated the feasibility of using pulsed laser ablation for SEI layer removal [51]. They utilized a Nd:YAG laser in the third harmonic with λ = 337 nm at an energy density of 55 mJ/cm^2 and a pulse width of 7 ns. Successful removal of the SEI without damaging the graphite active material was confirmed by SEM and Raman. **Figure 10(a)** and **(b)** provide a comparison of the graphite anode before and after 70 galvanostatic cycles, respectively. Before cycling, smooth graphite surfaces and sharp edges are apparent. Comparatively, the SEI forms after cycling, making the anode surface appear dull and coarse. With laser treatment, Zhang et al. were able to restore the smooth and sharp features of the graphite (**Figure 10(c)**), indicating the removal of the SEI layer. The Raman spectra given in **Figure 10(e)** confirm the regeneration of the graphite anode via similar I_D/I_G ratios (i.e., no change in graphite structural defects) for the pristine and laser-treated samples.

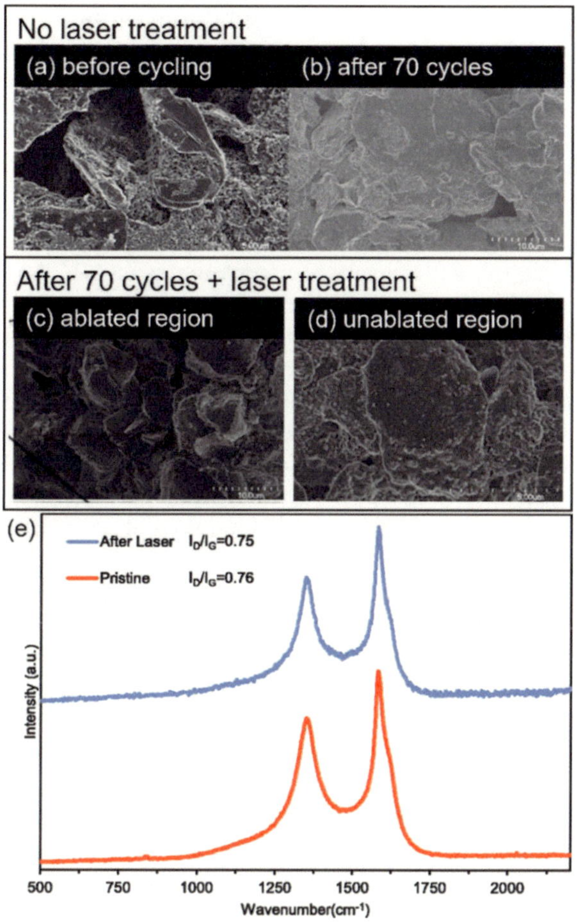

Figure 10.
SEM images of the graphite anode taken (a) before and (b) after 70 galvanostatic cycles. After cycling, the electrode was laser-treated to remove the SEI layer. SEM images show the (c) ablated and (d) unablated regions. (e) Raman spectra of the graphite electrode in the pristine condition (before cycling), compared to after cycling and laser-treatment. Reproduced with permission from [51]. Copyright (2017) Laser Institute of America.

4.3 Cutting solid electrolyte

Pulsed lasers are similarly being investigated for their role in processing solid ceramic electrolyte in the scaled and economic production of all-solid-state batteries. Solid ceramic electrolyte layers typically have thicknesses less than 100 μm, and they must remain defect-free after processing in order to prevent short circuiting between the cathode and anode [52]. Melt-quenching [53] and sintering [54] are the conventional methods used to produce low-porosity solid ceramic electrolyte, however, these often cause distortion and changes in geometry that have downstream consequences such as compromised contact with the adjacent electrodes. By applying machining to cut the solid electrolyte layers post-densification (i.e., remove distorted regions), precise geometries with improved mechanical properties can be restored. Laser ablation emerges as the most suitable machining method for processing solid ceramic electrolyte due to the hardness, brittleness and thin nature of the solid ceramic electrolyte. As a result, recent work has demonstrated the feasibility of using nano and picosecond lasers for cutting thin sheets of lithium aluminum titanium phosphate ($Li_{1+x}Al_xTi_{2-x}(PO_4)_3$, LATP) [55] and lithium aluminum germanium titanium phosphate ($Li_{1+x+3z}Al_x(Ti,Ge)_{2-x}Si_{3z}P_{3-z}O_{12}$, LAGTP) [56] solid electrolytes.

5. Pulsed lasers for chemical composition analysis

A final application of pulsed lasers in Li-ion battery research is the use of laser ablation inductively coupled plasma mass spectrometry (LA-ICP-MS) for chemical composition analysis of Li-ion batteries throughout their operating cycle. This technique is positioned as means of understanding the underlying mechanisms related to SEI formation and can help to inform optimization strategies. Other analysis techniques such as X-ray photoelectron spectroscopy (XPS) [57], time of flight-secondary ion mass spectrometry (Tof-SIMS) [58] and electron energy loss spectroscopy (EELS) [59] have been used to investigate the composition of SEI on Li-ion battery anodes. While these techniques provide insight into composition at depths resolved to the range of a few nanometers, they have limited lateral resolution. Conversely, the LA-ICP-MS technique provides a lateral spot resolution on the order of several tens of μm, enabling the facile analysis of entire electrode surfaces via rapid scanning.

As depicted in **Figure 11**, in LA-ICP-MS, a pulsed laser is used to ablate the surface of a sample. The particles generated by ablation are transported via argon carrier gas to an ICP plasma torch, where they are ionized. The ionized particles are then transferred to a mass spectrometer for elemental and isotopic analysis [60]. Currently, the most common applications of LA-ICP-MS are in geological research [61].

In 2017, Schwieters et al. first demonstrated proof-of-concept for the use of LA-ICP-MS to quantify the amount of lithium in SEI layers on aged graphite anodes [62]. Since lithium immobilized in the SEI contributes to decreased capacity, this work helps to advance the study and quantification of Li-ion battery health.

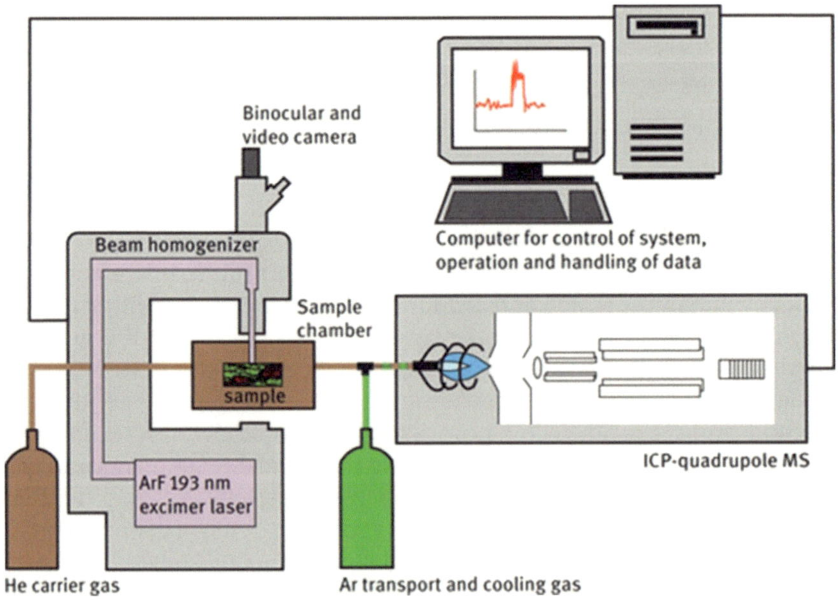

Figure 11.
Schematic of LA-ICP-MS. Reproduced with permission from [60]. Copyright (2017) De Gruyter.

Another degradative phenomenon occurring during cyclic aging of Li-ion batteries with transition metal oxide cathodes is the dissolution of transition metal cathode elements like Ni, Mn and Co and their re-deposition on the anode. Schwieters et al. applied LA-ICP-MS to investigate the influence of operating parameters on transition metal deposition [63]. They found that cells operating at higher voltages developed significantly more lithium and transition metal deposits, indicated increased SEI formation. This work demonstrates the promise of using LA-ICP-MS for compositional SEI analysis.

6. Conclusions

Pulsed laser ablation processes are scalable, environmentally sustainable and provide a high degree of machining precision. Moreover, machining in ultrafast pulses prevents thermal diffusion and permits robust material applicability. Consequently, pulsed lasers have been applied across many areas of Li-ion battery research, resulting in significant advancements. This chapter reviewed the use of pulsed lasers for nanoparticle generation, thin film deposition, machining and chemical composition analysis. In summary:

- Pulsed laser ablation in liquid was used to create nanoparticulate active material with high surface area, while laser ablation oxidation permitted integrated synthesis of nanoparticles directly on the current collector.
- Pulsed laser deposition was used to grow compositionally complex 2D thin films of cathode active materials in solid state microbatteries.
- Pulsed laser ablation was used to create 3D electrode architectures with improved diffusion kinetics and electrolyte wettability. Where Si-based anodes were used, laser texturing was used to accommodate volume expansions and reduce mechanical stress leading to failure.
- Pulsed laser ablation was used to regenerate the capacity of aged Li-ion batteries by removing the SEI layer from the anode surface.
- Pulsed lasers were used in the scaled cutting and processing of solid ceramic electrolyte.
- Laser ablation inductively coupled plasma mass spectrometry (LA-ICP-MS) was used to analyze SEI composition.

Overall, this chapter demonstrates the benefits realized by applying pulsed lasers for processing materials and components in Li-ion battery manufacturing and research. These methods should continue to be applied at the research-scale in order to develop next-generation batteries with improved performance. Additionally, future work should focus on scaling the developed ablation methods to ensure commercial viability in large-scale battery manufacturing processes.

Conflict of interest

Dongfang Yang is the Editor and Katherine Gibson is the Co-Editor of this IntechOpen book entitled "Pulsed Laser Processing of Materials."

Author details

Katherine Gibson* and Dongfang Yang
National Research Council Canada, London, ON, Canada

*Address all correspondence to: kategibson527@gmail.com

References

[1] Ritchie H. Cars, planes, trains: Where do CO_2 emissions from transport come from? Published online at OurWorldInData.org. 2020. Retrieved from: https://ourworldindata.org/co2-emissions-from-transport [Online Resource]

[2] Miao Y, Hynan P, von Jouanne A, Yokochi A. Current Li-ion battery technologies in electric vehicles and opportunities for advancements. Energies. 2019;**12**(6):1074

[3] Hannan MA, Hoque MM, Hussain A, Yusof Y, Ker PJ. State-of-the-art and energy management system of lithium-ion batteries in electric vehicle applications: Issues and recommendations. IEEE Access. 2018;**6**:19362-19378

[4] Yang Z, Zhu P, Ullah Z, Zheng S, Yu S, Zhu S, et al. Synchronous light harvesting and energy storing organic cathode material 1.4-dihydroxyanthraquinone for lithium-ion batteries. Chemical Engineering Journal. 2023;**468**:1-9

[5] Deng D. Li-ion batteries: Basics, progress, and challenges. Energy Science & Engineering. 2015;**3**(5):385-418

[6] Zhu P, Gastol D, Marshall J, Sommerville R, Goodship V, Kendrick E. A review of current collectors for lithium-ion batteries. Journal of Power Sources. 2021;**485**:229321

[7] Ashfold MNR, Claeyssens F, Fuge GM, Henley SJ. Pulsed laser ablation and deposition of thin films. Chemical Society Reviews. 2004;**33**:12-31

[8] Yan Z, Chrisey DB. Pulsed laser ablation in liquid for micro−/nanostructure generation. Journal of Photochemistry and Photobiology C: Photochemistry Reviews. 2012;**13**(3):204-223

[9] Tsuji T, Tatsuyama Y, Tsuji M, Ishida K, Okada S, Yamaki J. Preparation of $LiMn_2O_4$ nanoparticles for Li ion secondary batteries by laser ablation in water. Materials Letters. 2007;**61**(10):2062-2065

[10] Nowak AP, Lisowska-Oleksiak A, Siuzdak K, Sawczak M, Gazda M, Karczewski J, et al. Tin oxide nanoparticles from laser ablation encapsulated in a carbonaceous matrix—A negative electrode in lithium-ion battery applications. RSC Advances. 2015;**5**(102):84321-84327

[11] Alrefaee SH, Alkallas FH, Trabelsi ABG, Pashameah RA, Elsharkawy WB, Al-Ahmadi AN, et al. Laser assisted method for synthesis $Li_4Ti_5O_{12}$/polyether sulfone composite for lithium ion batteries anodic materials. Journal of Materials Research and Technology. 2023;**25**:440-450

[12] Luo X, Zhang H, Pan W, Gong J, Khalid B, Zhong M, et al. SiOx nanodandelion by laser ablation for anode of lithium-ion battery. Small. 2015;**11**(45):6009-6012

[13] Qiang W, Huanhuan H, Jian W, Zhurui S. Fabrication of SiOx ultra-fine nanoparticles by IR nanosecond laser ablation as anode materials for lithium ion battery. Applied Surface Science. 2017;**422**:155-161

[14] Liang P, Zhang H, Su Y, Huang Z, Wang C-A, Zhong A. In situ preparation of a binder-free nano-cotton-like CuO-Cu integrated anode on a current

collector by laser ablation oxidation for long cycle life Li-ion batteries. Journal of Materials Chemistry A. 2017;**5**(37):19781-19789

[15] Liang P, Zhang H, Pan B, Su Y, Wang C-A, Zhong M. Binder-free carbon-coated nanocotton transition metal oxides integrated anodes by laser surface ablation for lithium-ion batteries. Surface and Interface Analysis. 2019;**51**(8):874-881

[16] Su Y, Tong R-A, Zhang H, Liang P, Wang C-A, Zhong M. Defocused laser ablation process—A high-efficiency way to fabricate MoO_3-Mo integrative anode with excellent electrochemical performance for lithium ion batteries. Journal of Alloys and Compounds. 2019;**787**:295-300

[17] Cheung JT. History and fundamentals of pulsed laser deposition. In: Pulsed Laser Deposition of Thin Films. New York: John Wiley & Sons; 1994. pp. 1-22

[18] Antaya M, Dahn JR, Preston JS, Rossen E, Reimers JN. Preparation and characterization of $LiCoO_2$ thin films by laser ablation deposition. Journal of the Electrochemical Society. 1993;**140**(3):575-578

[19] Perkins J, Bahn CS, Parilla PA, McGraw JM, Fu ML, Duncan M, et al. $LiCoO_2$ and $LiCo_{1-x}Al_xO_2$ thin film cathodes grown by pulsed laser ablation. Journal of Power Sources. 1999;**81**:675-679

[20] O'Mahony D, Lunney J, Dumont T, Canulescu S, Lippert T, Wokaun A. Laser-produced plasma ion characteristics in laser ablation lithium manganate. Applied Surface Science. 2007;**254**(4):811-815

[21] Julien C, Haro-Poniatowski E, Camacho-Lopez MA, Escobar-Alarcon L, Jimenez-Jarquin J. Growth of $LiMn_2O_4$ thin films by pulsed-laser deposition and their electrochemical properties in lithium microbatteries. Materials Science and Engineering: B. 2000;**72**(1):36-46

[22] Rougier A, Striebel K, Wen SJ, Richardson TJ, Reade RP, Cairns EJ. Characterization of pulsed laser-deposited $LiMn_2O_4$ thin films for rechargeable lithium batteries. Applied Surface Science. 1998;**134**:107-115

[23] Morecette M, Barboux P, Perriere J, Brousse T, Traverse A, Boilot JP. Non-stoichiometry in $LiMn_2O_4$ thin films by laser ablation. Solid State Ionics. 2001;**138**:213-219

[24] Teranishi T, Yoshikawa Y, Leblanc-Lavoie J, Delegan N, Ka I, Kishimoto A, et al. Capacity retention improvement of $LiCoO_2$ cathodes via their laser-ablation-based nanodecoration by BaTiO3 nanoparticles. Journal of Applied Physics. 2022;**131**(12):124105

[25] Wang Y, Qin Q-Z. A nanocrystalline NiO thin-film electrode prepared by pulsed laser ablation for Li-ion batteries. Journal of the Electrochemical Society. 2002;**149**(7):A873-A878

[26] Curcio M, De Bonis A, Brutti S, Santagata A, Teghil R. Pulsed laser deposition of thin films of TiO_2 for Li-ion batteries. Applied Surface Science Advances. 2021;**4**:100090

[27] Pfenninger R, Afyon S, Garbayo I, Struzik M, Rupp J. Lithium titanate anode thin films for Li-ion solid state battery based on garnets. Advanced Functional Materials. 2018;**28**(21):1800879

[28] Park MS, Wang GX, Liu HK, Dou SX. Electrochemical properties of Si thin film prepared by pulsed laser deposition for lithium ion

micro-batteries. Electrochimica Acta. 2006;**51**(25):5246-5249

[29] Naoaki K. Characterization of thin-film lithium batteries with stable thin-film Li_3PO_4 solid electrolytes fabricated by ArF excimer laser deposition. Journal of the Electrochemical Society. 2010;**157**(4):A521

[30] Zhao S, Qin Q. LiVSiO thin film electrolyte for all-solid-state Li-ion battery. Journal of Power Sources. 2003;**122**(2):174-180

[31] Reinacher J, Berendts S, Janek J. Preparation and electrical properties of garnet-type $Li_6BaLa_2Ta_2O_{12}$ lithium solid electrolyte thin films prepared by pulsed laser deposition. Solid State Ionics. 2014;**258**:1-7

[32] Kuwata N, Kawamura J, Toribami K, Hattori T, Sata N. Thin-film lithium-ion battery with amorphous solid electrolyte fabricated by pulsed laser deposition. Electrochemistry Communications. 2004;**6**:417-421

[33] Gower MC. Industrial applications of laser micromachining. Optics Express. 2000;**7**(2):56-67

[34] Zhu P, Seifert HJ, Pfleging W. The ultrafast laser ablation of $Li(Ni_{0.6}Mn_{0.2}Co_{0.2})O_2$ electrodes with high mass loading. Applied Sciences. 2019;**9**(19):4067

[35] Proll J, Kim H, Pique A, Seifert HJ, Pfleging W. Laser-printing and femtosecond-laser structuring of $LiMn_2O_4$ composite cathodes for Li-ion microbatteries. Journal of Power Sources. 2014;**255**:116-124

[36] Proll J, Kohler R, Torge M, Ulrich S, Ziebert C, Bruns M, et al. Laser microstructuring and annealing processes for lithium manganese oxide cathodes. Applied Surface Science. 2011;**257**:9968-9976

[37] Mangang M, Seifert HJ, Pfleging W. Influence of laser pulse duration on the electrochemical performance of laser structures $LiFePO_4$ composite electrodes. Journal of Power Sources. 2016;**304**:24-32

[38] Zhu P, Ebert B, Smyrek P, Pfleging W. The Impact of Structural Pattern Types on the Electrochemical Performance of Ultra-Thick NMC 622 Electrodes for Lithium-Ion Batteries. 2024;**10**(2):1-21

[39] Baek G, Choi T-U, Kwon J-D, Ha J-H, Lee S-J, Lee SG, et al. Periodically aligned channels in $Li[Ni_{0.5}Co_{0.2}Mn_{0.3}]O_2$ cathodes designed by laser ablation for high power Li ion batteries. Journal of Energy Storage. 2022;**50**:104551

[40] Peethan A, Unnikrishnan VK, Chidangil S, George SD. Laser-assisted tailoring of surface wettability—Fundamentals and applications. In: Progress in Adhesion and Adhesives. Beverly, MA, USA: Scrivener Publishing LLC; 2020

[41] Kleefoot M-J, Enderle S, Sandherr J, Bolsinger M, Maischik T, Simon N, et al. Enhancement of the wettability of graphite-based lithium-ion battery anodes by selective laser surface modification using low energy nanosecond pulses. The International Journal of Advanced Manufacturing Technology. 2022;**118**:1987-1997

[42] Zheng Y, Seifert HJ, Shi H, Zhang Y, Kubel C, Pfleging W. 3D silicon/graphite composite electrodes for high-energy lithium-ion batteries. Electrochimica Acta. 2019;**317**:502-508

[43] Meyer A, Sterzi Y, Pfleging W. High repetition ultrafast laser ablation of graphite and silicon/graphite

composite electrodes for lithium-ion batteries. Journal of Laser Applications. 2023;**35**(4):042036

[44] Tancin RJ, Sular-Kern DB, Usseglio-Viretta FL, Finegan DP, Tremolet de Villers BJ. Characterization of ultrafast-laser ablation of micro-structures in Li-ion battery anode and cathode materials: Morphology, rate, and efficiency. Journal of Power Sources. 2024;**596**:1-13

[45] Zhang G, Hua X, Huang Y, Zhang Y, Fang L, Shen C, et al. Investigation on mechanism of oxide removal and plasma behaviour during laser cleaning on aluminum alloy. Applied Surface Science. 2020;**506**:144666

[46] Bradley B. 8 lessons about EV battery health from 6,300 electric cars. Published online at Electrek.co. 2019. Retrieved from: https://electrek.co/2019/12/14/8-lessons-about-ev-battery-health-from-6300-electric-cars [Online Resource] [Accessed: February 1, 2024]

[47] Saxena S, Le Floch C, MacDonald J, Moura S. Quantifying EV battery end-of-life through analysis of travel needs with vehicle powertrain models. Journal of Power Sources. 2015;**282**:265-276

[48] Aurbach D, Levi MD, Levi E, Schechter A. Failure and stabilization mechanisms of graphite electrodes. The Journal of Physical Chemistry B. 1997;**101**(12):2195-2206

[49] Edstrom K, Herstedt M, Abraham DP. A new look at the solid electrolyte interphase on graphite anodes in Li-ion batteries. Journal of Power Sources. 2006;**153**(2):380-384

[50] Abraham DP, Knuth JL, Dees DW, Bloom I, Christophersen JP. Performance degradation of high-power lithium-ion cells—Electrochemistry of harvested electrodes. Journal of Power Sources. 2007;**170**(2):465-475

[51] Zhang Y, Zhang Y, Liu Z, Guan D, Wang F, Meyers V, et al. Laser ablation on lithium-ion battery electrode solid electrolyte interface removal. Journal of Laser Applications. 2017;**29**(4):042002

[52] Knauth P. Inorganic solid Li ion conductors: An overview. Solid State Ionics. 2009;**180**:911-916

[53] Rohde M, Cui Y, Ziebert C, Seifert HJ. Thermophysical properties of lithium aluminum germanium phosphate with different compositions. International Journal of Thermophysics. 2020;**41**:1-13

[54] Dashjav E, Gellert M, Yan G, Gruner D, Kaiser N, Spannenberger S, et al. Microstructure, ionic conductivity and mechanical properties of tape-cast Li1.5Al0.5Ti1.5P3O12 electrolyte sheets. Journal of the European Ceramic Society. 2020;**40**(5):1975-1982

[55] Kriegler J, Jaimez-Farnham E, Hille L, Dashjav E, Zaeh MF. Pulsed laser ablation of a ceramic electrolyte for all-solid-state batteries. Procedia CIRP. 2022;**111**:800-805

[56] Kriegler J, Hille L, Oehler A, Chaja M, Zaeh MF. Scaling up picosecond laser ablation of a LATGP-type glass-ceramic solid electrolyte for all-solid-state battery production. Journal of Manufacturing Processes. 2023;**106**:188-201

[57] Niehoff P, Passerini S, Winter M. Interface investigations of a commercial lithium ion battery graphite anode material by sputter depth profile X-ray photoelectron spectroscopy. Langmuir. 2013;**29**(19):5806-5816

[58] Gachot G, Grugeon S, Armand M, Pilard S, Guenot P, Tarascon J-M, et al.

Deciphering the multi-step degradation mechanisms of carbonate-based electrolyte in Li batteries. Journal of Power Sources. 2008;**178**:409-421

[59] Wang F, Graetz J, Moreno MS, Ma C, Wu L, Volkov V, et al. Chemical distribution and bonding of lithium in intercalated graphite: Identification with optimized electron energy loss spectroscopy. ACS Nano. 2011;**5**(2):1190-1197

[60] Sindern S. Analysis of rare earth elements in rock and mineral samples by ICP-MS and LA-ICP-MS. Physical Sciences Reviews. 2017;**2**(2):1-14

[61] Rusk B. Laser ablation ICP-MS in the Earth sciences: Current practices and outstanding issues. Economic Geology. 2009;**104**(4):601-602

[62] Schwieters T, Evertz M, Mense M, Winter M, Nowak S. Lithium loss in the solid electrolyte interphase: Lithium quantification of aged lithium ion battery graphite electrodes by means of laser ablation inductively coupled plasma mass spectrometry and inductively coupled plasma optical emission spectroscopy. Journal of Power Sources. 2017;**356**:47-55

[63] Schwieters T, Evertz M, Fengler A, Borner M, Dagger T, Stenzel Y, et al. Visualizing elemental deposition patters on carbonaceous anodes from lithium ion batteries: A laser ablation-inductively coupled plasma-mass spectrometry study on factor influencing the deposition of lithium, nickel, manganese and cobalt after dissolution. Journal of Power Sources. 2018;**380**:194-201

Chapter 2

Fundamentals of Ultrashort Pulse Laser Interactions: Mechanisms, Material Responses, and the Genesis of LIPSS

Hardik Vaghasiya and Paul-Tiberiu Miclea

Abstract

In recent years, ultrashort pulse laser-material processing has gained significant attention due to its broad applications across nearly all manufacturing sectors. This chapter delves into the foundational aspects of the ultrashort pulse laser-material interaction and elucidates the intricacies of the underlying ablation mechanisms. Due to peculiarities between the metal energy absorption in contrast to the semiconductor or dielectric, the first section provides an in-depth exploration of laser-material dynamics, emphasizing the unique responses of various substrates under ultrashort pulse irradiation. A theoretical analysis of ultrashort laser-matter interaction can be represented by the two-temperature model, which describes the temperature of the electron or carrier and lattice in non-equilibrium conditions when ultrashort laser pulses are applied. As the narrative progresses, the spotlight shifts to one of the most interesting phenomena associated with these interactions: the formation of Laser-Induced Periodic Surface Structures (LIPSS). The second section unravels the genesis and evolution of LIPSS, demystifying LIPSS formation mechanisms and the pivotal role played by the ultrashort pulse duration.

Keywords: ultrashort pulse laser, two-temperature model, carrier density two-temperature model, LIPSS formation, ablation mechanism

1. Introduction

The study of ultrashort pulse laser interactions with materials has become increasingly significant in recent years, driven by its wide-ranging applications in various manufacturing sectors [1, 2]. To optimize laser processes and improve efficiency and quality, it is essential to have a deep understanding of the underlying physics and material-specific responses when laser pulses interact with material. Moreover, this knowledge enables precise customization of laser parameters for applications like micro and nanofabrication and surface modification. Fundamental studies also drive innovation, helping develop new techniques and pushing the boundaries of what ultrashort pulse lasers can achieve.

Figure 1 demonstrates the laser processing workflow, divided into four distinct parts, each playing a crucial role in the precision and efficiency of the laser-based manufacturing process. Single laser pulse width cannot be done with desired micromachining with all materials; a choice of laser (nanosecond, picosecond, and femtosecond) and their parameter are essential for the process's quality, throughput, and cost-effectiveness. In nanosecond laser-material processing, the precision and quality of micromachining are not as high as those achieved in femtosecond and picosecond laser micromachining, primarily due to the presence of a heat-affected zone (HAZ) and re-solidification of the ablated feature. Lasers with their excellent beam quality promised noticeable advantages and improvements in high precision and material processing at the micron scale. Thus, a new generation of laser beam sources with ultrashort pulses and high repetition rates has been used in vast industrial applications, and it offers excellent improvements, especially in the fields of laser processing and micro-manufacturing. Ultrashort laser pulses in the range of picoseconds and femtoseconds have high pulse intensities and enable high-resolution, high-quality micromachining of almost all solid materials with minimal thermal and mechanical damage and structure sizes down to the submicron range. It is an excellent and universal tool for very fine, direct ablative microstructuring of solid materials. It has been demonstrated that ultrashort pulses bear the potential for precise micromachining (laterally and vertically) in transparent dielectrics as well as metal and semiconductors [3, 4].

The first section provides a fundamental aspect of ultrashort laser-material processing, driven by the peculiarities inherent in the energy absorption of metals compared to semiconductors or dielectrics. This section offers an insight into laser-material dynamics and the response of various substrates while irradiation of ultrashort laser pulses. A two-temperature model is used to analyze the intricacies of ultrashort laser-matter interactions, shedding light on the non-equilibrium conditions that arise within the electron or carrier and lattice temperature. The second section is devoted to the description of the LIPSS and the potential mechanisms of their formation.

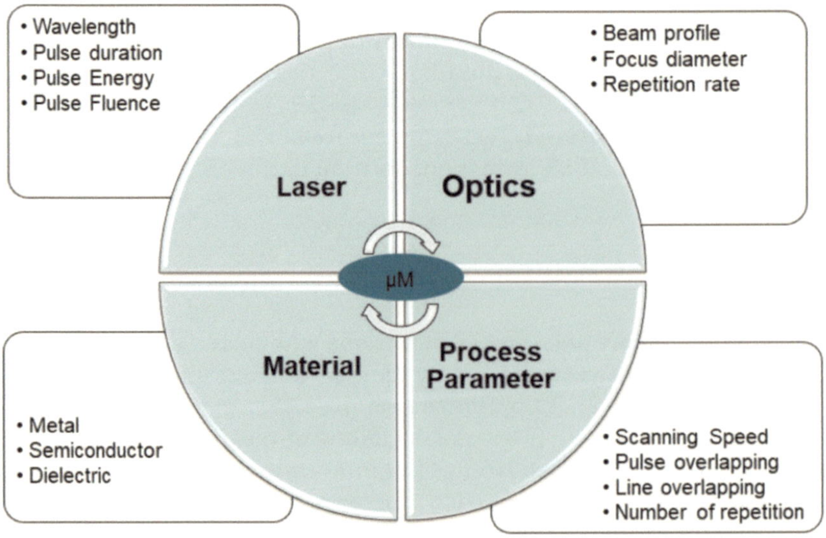

Figure 1.
Comprehensive illustration showcasing laser-material processing.

2. Two-temperature model (TTM)

In the realm of ultrashort laser-material interaction, the material response can be described as a result of elevated temperatures. Therefore, it is important to understand the flow of heat inside a material. The two-temperature model represents the temporal and spatial evolution of the temperature inside a material.

2.1 Metal

Figure 2 depicts a schematic diagram of ultrashort laser pulses acting on the metal surfaces. It is divided into five parts—absorption, heating, energy transfer, thermomechanical response of the material, and ablation. When the laser pulses irradiate the substrate or material, energy transfers from the laser pulses to the material; this energy is absorbed by free electrons, which increases the temperature of the electron (absorption). At room temperature, the lattice's and electron's temperatures are in thermal equilibrium. Once an electron gets energy from laser pulses, an electron-electron collision occurs. As a result, there is a rapid increase in the electron temperature (heating), and these energetic electron transfers their energy to the lattice by electron-lattice collision (energy transfer). Numerous energetic electrons are present during pulsed laser irradiation of the material, requiring multiple electron-lattice collisions to transfer a significant amount of energy. Due to the transfer of energy from electron to lattice system, heat diffusion in material or substrate occurs (thermomechanical response), and material is ablated (ablation). The evolution of electron and lattice temperatures over time is determined by the two-temperature model (TTM) and governing equations of two-temperatures are given below [5].

$$C_e \frac{\partial T_e}{\partial t} = \nabla(k_e \nabla T_e) - k(T_e - T_i) + Q_{total} \tag{1}$$

$$C_l \frac{\partial T_l}{\partial t} = \nabla(k_l \nabla T_l) - G(T_e - T_l) \tag{2}$$

where T_e and T_l are the temperature of the electron and lattice, k_l and k_e are the thermal conductivity of the lattice and electron, C_e and C_l are the heat capacity of the

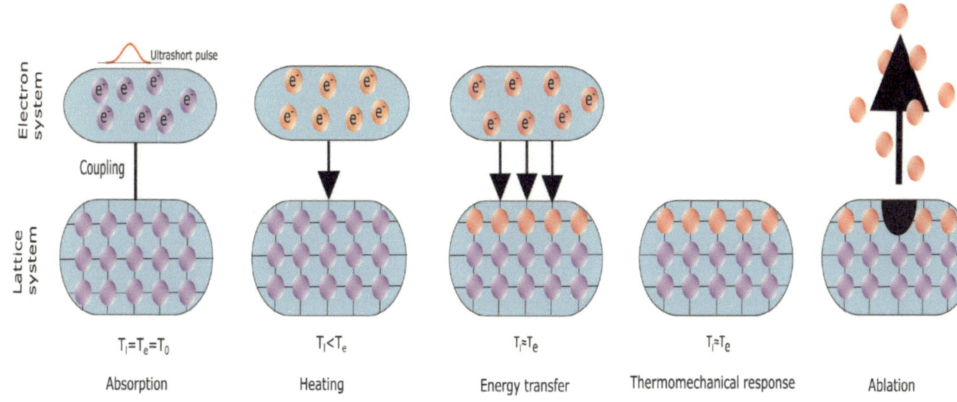

Figure 2.
Schematic diagram of ultrashort pulse laser interaction with metal. Where T_l and T_e are temperatures of lattice and electron and T_0 is the room temperature.

electron and lattice. Q_{total} represents the absorbed laser heating source and G is the electron-lattice coupling. During the ultrafast heating of the material, the temperature of electrons and lattices rises so that temperature-dependent materials' properties have a pivotal role during the process.

As can be seen in **Figure 2**, two mechanisms of energy transfer, from laser to electron and from electron to lattice, involve two different temporal time scales driven by the electron relaxation time (τ_e) and electron-lattice relaxation time (τ_{ep}). In metal, the electron cooling time is around a hundred femtoseconds, while the electron-lattice relaxation time corresponds to several picoseconds. The electron-lattice coupling factor represents the energy exchange rate between the two subsystems (electron and lattice), and research groups suggest that it depends on the electron temperature, which is linear or nonlinear [6, 7]. Drastic changes in electron temperature and lattice temperature during and after irradiation of ultrashort laser pulses create a decisive influence on the material's optical properties, such as absorptivity and reflectivity. The laser heat source can be delineated by the following equation, and it gives the absorption of laser energy by free electron [7].

$$Q_{total} = (1-R) \times F \times \alpha \times \exp\left(-2\frac{r^2}{\omega_0^2}\right) \times \exp\left(-4\ln 2\frac{(t-t_0)^2}{\tau^2}\right) \times \exp(\alpha z) \quad (3)$$

where R representing the surface reflectively, F is the laser fluence, τ is the pulse width, ω_0 is the beam radius at the laser focal plane and the pulse peak arrives at t_0. R and α are the reflectivity and absorption coefficient of the material which have a significant influence on the laser energy distribution on the material transition. The reflectivity and absorption coefficient can be derived in terms of the dielectric constant ε_1 and ε_2. Generally, the Drude model is used to determine the dielectric constant and intraband absorption in the metal. The dielectric constant based on the Drude model is described by [8]:

$$\varepsilon_D = \varepsilon_\infty - \frac{\omega_p^2}{\omega(\omega + i\gamma_D)} = \varepsilon_1 + i\varepsilon_2 \quad (4)$$

where ε_∞ is a the dielectric constant of the material and ω_p is the plasma frequency, ω is the laser frequency, and γ_D is the damping coefficient that is equal to the reciprocal of electron relaxation time τ_e. It can be expressed as follows

$$\tau_e = \frac{1}{A_e T_e^2 + B_l T_l} \quad (5)$$

where A_e and B_l are the material constant for electron relaxation time. Based on the Drude model, refractive index $n = \sqrt{\frac{\varepsilon_1 + \sqrt{\varepsilon_1^2 + \varepsilon_2^2}}{2}}$ and extinction coefficient $k = \sqrt{\frac{-\varepsilon_1 + \sqrt{\varepsilon_1^2 + \varepsilon_2^2}}{2}}$ can be determined. According to the Fresnel equation, the reflectivity(R) and absorption coefficient (α) is obtained by [9].

$$R = \frac{(n-1)^2 + \kappa^2}{(n+1)^2 + \kappa^2} \quad (6)$$

$$\alpha = \frac{4\pi\kappa}{\lambda} \quad (7)$$

In metals, the absorption of laser light becomes more intricate due to the involvement of interband absorption mechanisms. For instance, in polyvalent materials like aluminum, nearly parallel bands lead to an increased absorption at certain laser frequencies. On the other hand, noble metals present a challenge due to their complex energy band structure. When sufficient high photon energies are applied to the noble metal, d-electrons may be excited into the s-band above the Fermi level, which contradicts our predictions made using the simply Drude model. Hence, in order to provide an accurate description of such phenomena, more advanced theoretical approaches are required. In noble metal, experimental work has shown that the effective penetration depth of the laser pulse in the material is larger than the optical penetration depth because of the ballistically moving electrons, and this is called the ballistic effect. Considering the ballistic effect, the free carrier absorption coefficient can be derived as follows:

$$\alpha_{abs} = \frac{1}{\delta + \delta_{\text{ball}}} \tag{8}$$

With δ denoting the optical penetration depth, which is the inverse of the absorption coefficient (α). The δ_{ball} is the ballistic range because of the ballistic motion and diffusion of the electrons. The previous research work demonstrated a ballistic range of about 15 nm for copper [10], 50 nm [10] for silver, and 100 nm for gold [11]. Researchers used advanced diagnostic techniques such as ultrafast pump-probe spectroscopy and time-resolved electron diffraction to measure electron and lattice temperatures in real time as well as find the basic length. These experiments not only confirm the theoretical predictions of the two-temperature model but also provide valuable insights into the transient nature of metal response during ultrashort laser ablation. By studying the evolution of the electron and lattice temperature, researchers can predict material responses, optimize laser parameters, and advance the precision and efficiency of material processing.

A numerical simulation based on the TTM model for the gold sample is shown in **Figure 3** [12]. The electron temperature increases quickly and reaches a peak value of

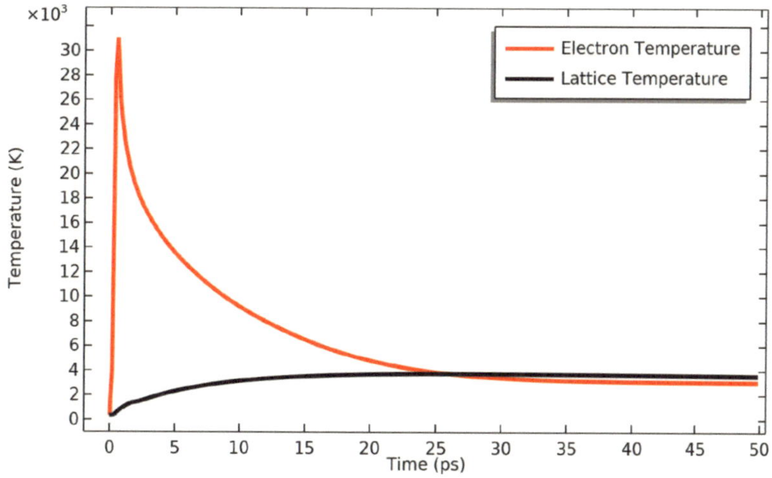

Figure 3.
Time evolution of electron temperature () and lattice temperature (T_l) of the gold sample irradiated by the 180 fs laser with the wavelength of 515 nm at fluence of F = 0.50 J/cm².

around 31,200 K at t = 2 ps, as shown in **Figure 3**. The lattice temperature begins to increase quickly after t = 1 ps due to the transferred energy from the electrons to lattices, and then reaches a maximum value of 3880 K at t = 26 ps. Afterward, the temperature of the electron and lattice is almost the same.

2.2 Semiconductors and dielectrics

Until now, only laser ablation with material that has free electrons (metal) has been discussed in two-temperature models, but materials like semiconductors and dielectric material ablation processes differ as compared to metals. During ultrashort pulse irradiation in semiconductors and dielectric materials, the crucial processes of electron-hole pair generation and relaxation are facilitated through multiphoton ionization and tunneling effects. While ultrashort laser pulses interact with high band gap materials, the pulse duration is shorter than the carrier-lattice interaction time. This leads to a non-equilibrium state between the carrier and lattice systems throughout the irradiation. The energy transfer within the material can be accurately described by the carrier density, denoted as n-TTM [13]. **Figure 4** shows a schematic diagram of ultrashort laser pulses acting on the surface of large band gap materials. It is divided into five parts: absorption, heating, energy transfer, thermomechanical response of the material, and ablation. When the laser pulses irradiate the semiconductor, energy is transferred from the laser pulses to the material. This energy is absorbed by electrons and excited from the valence band to the conduction band via single-photon or multiphoton absorption depending on the laser energy and bandgap of the material, resulting in the creation of electron-hole pairs. In dielectric materials, electron-hole pairs are generated through two key processes: photoionization and avalanche ionization. Photoionization involves electrons moving from the valence to the conduction band via multiphoton or tunneling processes. Multiphotoionization or tunneling ionization mainly occurs at a high intensity and power density. When a single photon does not have enough energy to overcome a binding energy gap between the valance band and conduction band, a multiphoton is necessary to overcome this binding energy gap. Alternatively, when the electric field induced by the laser effectively suppresses the Coulomb wall, electrons can liberate themselves by tunneling from the valence to the conduction band—this phenomenon is termed tunneling ionization.

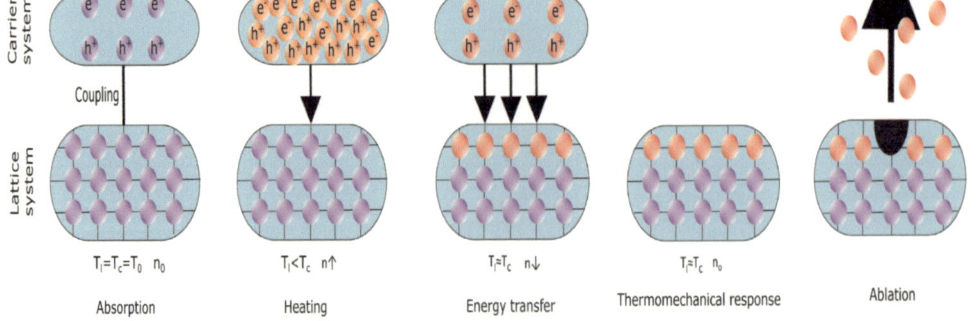

Figure 4.
Schematic diagram of ultrashort pulse laser interaction with high band gap materials. Where T_l and T_c are the temperature of the lattice and carrier, and n_0 is the initial carrier density of a material.

Moreover, under certain conditions, multiphotoionization can be further intensified by the irradiating laser pulse, leading to the generation of multiple free electrons in the valence band. This intricate process, known as avalanche ionization, plays a pivotal role in plasma formation during ultrashort laser ablation of dielectric materials. New electron-hole pairs are generated by the energetic charge carrier via the impact ionization process, and the carrier density increases with the carrier temperature (heating). Afterward, these free electrons and holes recombine mainly via the Auger recombination process and transfer the excess energy to another free electron-hole pair. Simultaneously, the carrier system couples to the lattice system and transfers the energy from the carrier to the lattice until thermal equilibrium (energy transfer). Because of the energy transfer from the carrier to the lattice, the top surface of the material reaches the other state (thermomechanical response). Eventually, the temperature of the top surface reaches the vaporization temperature of the material, causing ablation (ablation).

To understand the physical phenomenon of ultrashort laser light absorption in high bandgap material, one has to take into account the changes in the density of the free electron in the conduction band of material. The formation of electron-hole pairs (carrier) during irradiation is critical, as it significantly influences the absorption dynamics and subsequent material responses in high band gap materials. The rate of the change of carrier density due to laser excitation processes can be expressed as follows [8]:

$$\frac{\partial n}{\partial t} = \frac{\alpha I}{h\nu} + \frac{\beta I^2}{h\nu} \tag{9}$$

Impact ionization and auger recombination processes significantly influence the generation of carrier density in semiconductors. By including them, the generation rate of the carrier density in semiconductors can be defined as follows [13]:

$$\frac{\partial n}{\partial t} = \nabla(D_0 \nabla n) - \gamma n^3 + \eta n + \frac{\alpha_a I}{h\nu} + \frac{\beta I^2}{h\nu} \tag{10}$$

where n is the number of carrier density, α_a is the one-photon absorption coefficient, β is the two-photon absorption coefficient, I is the laser intensity, h is the Planck constant, ν is the photon frequency, γ is the auger recombination coefficient, η is the impact ionization coefficient, and D_0 is the ambipolar diffusivity which represents the electron-hole pair mobility and transport from a valence band to a conduction band.

In wide bandgap dielectrics, the primary mechanisms involved in generating free carriers during ultrashort laser pulse irradiation are multiphoton and avalanche ionization. The process that takes place during ultrafast laser ablation of dielectric material, as represented by the Keldysh parameter (γ_K), includes multiphotonionization, tunneling ionization, and avalanche ionization [14, 15].

$$\gamma_K = \frac{\omega_L \sqrt{m_r \varepsilon_{gap}}}{eE_L} \tag{11}$$

where e signifies the elementary (positive) charge, m_r refers to the reduced mass of electrons and holes, E_l represents the electric field of the laser wave, and ε_{gap} is the band gap of dielectric materials or the ionization potential of individual atoms or molecules. Using the Keldysh parameter, the transition between two ionization (multiphoton ionization and tunneling ionization) can be estimated. There are three distinct cases based on the value of (γ_K). When $\gamma_K > 1$, multiphoton ionization prevails, signifying that the tunneling time of the electron is greater than the periodic oscillation of the laser. On the other hand, when $\gamma_K) < 1$, tunnel ionization dominates under very high electric fields and low frequencies. When the Keldysh parameter equals 1, both multiphoton and tunnel ionization processes contribute to the electron excitation.

The generation of carrier density due to the photoionization and impact ionization in dielectric material can be described by [16].

$$\frac{\partial n}{\partial t} = \sigma_k I^k + \delta I n \tag{12}$$

where σ_k is the multiphoton ionization cross section, δ is the the avalanche coefficient, and k is the number of photons required for multiphoton ionization process. This equation provides a straightforward approach to estimating experimental conditions, considering non-uniform carrier density profiles in the surface of dielectric material and variations in optical response. The excitation process is characterized by two stages: initial multiphoton ionization generating free electrons, followed by avalanche ionization once a critical carrier density is reached. At high fluence, tunneling ionization may prevail over multiphoton ionization, as indicated by the earlier mention of the Keldysh parameter.

The excitation of the free electron density mainly depends upon the ultrashort laser pulse intensity. In wide band gap materials, it becomes essential to consider both one-photon absorptions and two-photon absorption across the band gap. The significance of three-photon absorption may also vary depending on photon energy, but it can be neglected as compared to single-photon absorptions and two-photon absorption. The attenuation of laser intensity due to single and two-photon absorption, as well as free carrier absorption, is described by the following expression [13]:

$$\frac{\partial I(z,t,r)}{\partial z} = -(\alpha_a + \alpha_{\mathrm{FCA}})I - \beta I^2. \tag{13}$$

where α_{FCA} is the free carrier absorption. α_{FCA} can be associated with Drude absorption and can be determined with Eqs. (4) and (7). Moreover, the dielectric function ε may comprise more terms than the simplified Eq. (4) typically used for describing excitation in high band gap materials. It is essential to note that the dielectric function ε may comprise more terms than the simplified Eq. (4) typically used for describing excited semiconductors [17, 18]. Most of the time, the two-photon absorption coefficient is neglected because the photon energy is greater than the bandgap of material. Assuming a negligible two-photon absorption coefficient, the Lambert-Beer law can be applied, where $a_{\mathrm{abs}} = \alpha_a + \alpha_{\mathrm{fca}}$.

The laser beam is Gaussian, and the temporal and spatial evolution of the laser intensity at the top surface (z = 0) can be expressed as [13]:

$$I_0(r,t,z=0) = \sqrt{\frac{4\ln 2}{\Pi}}\left(\frac{F}{\tau_p}\right)\exp\left(-2\frac{r^2}{w_0^2}\right) \times (1-\mathrm{R})\exp\left(-4\ln 2\frac{(t-t_0)^2}{\tau_p^2}\right) \tag{14}$$

For the carrier energy balance equation, the spatial and temporal evolution of carrier temperature and lattice temperature can be defined as follows [13]:

$$C_c \frac{\partial T_c}{\partial t} = \nabla(k_c \nabla T_c) - k(T_c - T_l) + (\alpha + \beta I + \Theta n)I - \frac{\partial n}{\partial t}(E_g + 3K_B T_c) - n\left(\frac{\partial E_g}{\partial n}\frac{\partial n}{\partial t} + \frac{\partial E_g}{\partial T_c}\frac{\partial T_c}{\partial t}\right) \quad (15)$$

$$C_l \frac{\partial T_l}{\partial t} = \nabla(k_l \nabla T_l) + G(T_c - T_l) \quad (16)$$

where T_c and T_l are the temperatures of carrier and lattice, respectively, k_c and k_l are the thermal conductivity of the carrier and lattice, respectively, C_c and C_l are the heat capacity of the carrier and lattice, respectively, K_b is the Boltzmann constant, t is the time, Θ is the carrier absorption coefficient, E_g is the bandgap of material, G is a coupling constant.

Figure 5 shows the evolution of the carrier density, carrier temperature, and lattice temperature for femtosecond laser-silicon interaction [12]. The carrier temperature and lattice temperature rise as free carriers are generated. It is interesting to note that the peak carrier temperature occurs at t = 0.3 ps and 4600 K, while the peak carrier density occurs at 1.2 ps around $0.29 \times 10^{27} m^{-3}$, much later than the carrier temperature. This is due to the fact that the heat source of the carrier temperature and the energy change need that time to uplift carrier density. The carrier number density increases dramatically as time passes due to laser light excitation of electrons from the valance band to the conduction band via single-photon absorption, as shown by numerical results in **Figure 5**. The Auger recombination process causes it to decline after that. Simultaneously, due to energy transfer from the carrier to the lattice, the carrier temperature continues to fall as time passes after reaching a peak. Through the process of electron-lattice energy relaxation, the carrier and lattice temperatures reach equilibrium at 8 ps as time passes.

In recent studies investigating the non-equilibrium energy transfer dynamics in matter under ultrashort laser irradiation, researchers have employed advanced models to elucidate the complex interplay of energy absorption and thermal relaxation processes. For that purpose, two-temperature models have been further refined and

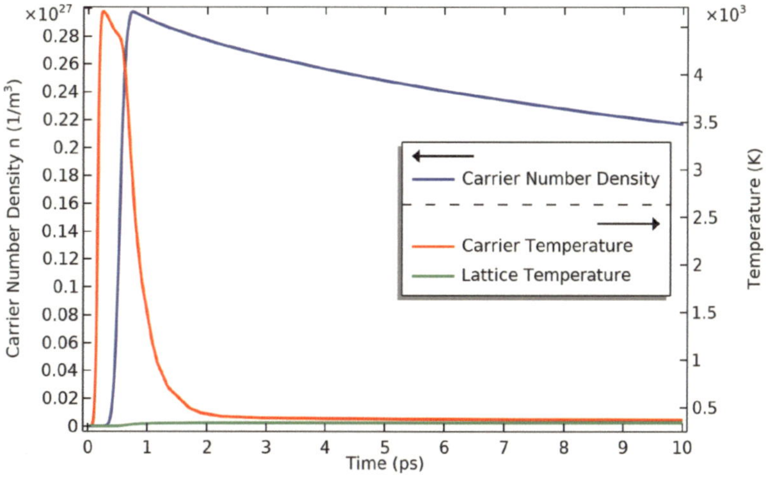

Figure 5.
Time evolution of carrier density(n), carrier temperature (T_c), and lattice temperature (T_l) of silicon sample irradiated by the 180 fs laser with the wavelength of 515 nm at fluence of 0.020 J/cm^2.

extended through various modifications, including considerations of parabolic, hyperbolic, dual-hyperbolic, and temperature-dependent optical properties [19–21]. Additionally, the development of alternative models, such as the Hyperbolic Two-Temperature Model (HTTM) and the Nonlocal Two-Temperature Model (NTTM), has provided deeper insights into energy transfer dynamics, incorporating relaxation time and space nonlocal effects based on extended irreversible thermodynamics principles [22]. Moreover, several research groups have combined the molecular dynamics (MD) simulation with TTM to investigate ultrashort laser-matter interactions, providing detailed atomic-level insights into morphological changes induced by laser irradiation [23, 24].

3. Ablation mechanism

Laser ablation occurs through a series of complex physical processes. When a high-intensity laser beam interacts with a solid material, several phenomena take place, such as the absorption of laser energy, heating, and melting, depending upon the type of laser and material. To disintegrate some amount of matter from a substrate, the material should undergo some change in phase, such as the vaporization of material, phase change, or plasma formation. **Figure 6** illustrates the phenomena and processes occurring during the laser irradiation of a solid. In nanosecond laser ablation, the pulse duration is relatively long so that it does not directly ionize or excite electrons from the material. Instead, the longer pulse duration allows for gradual energy absorption by the material, leading to thermal processes such as vaporization and melting, which

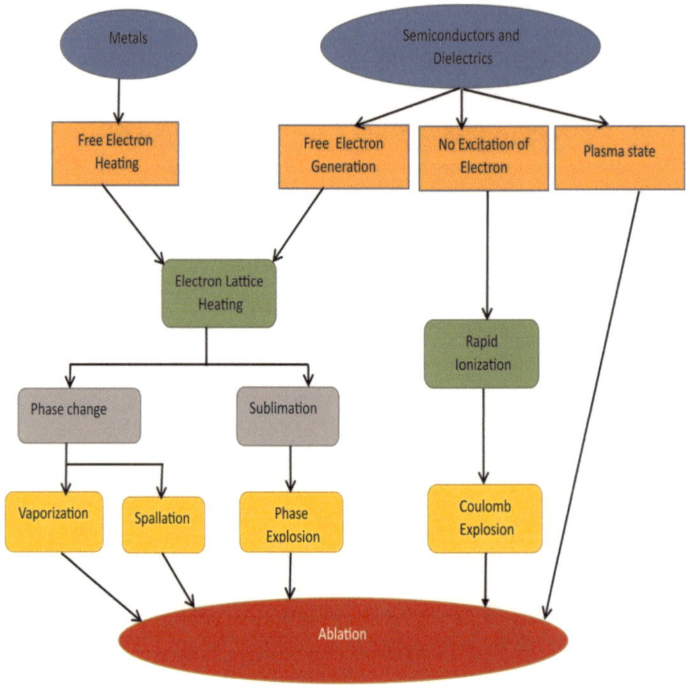

Figure 6.
Laser ablation dynamics in metal, semiconductor, and dielectric.

eventually result in material removal. Sometimes boiling can also occur, but only when the vapor pressure of the liquid phase exceeds the ambient pressure [25].

Upon the ultrashort laser irradiation, the strongly excited electrons can be created within a few femtoseconds. These excited electrons interact with the lattice on time-scales ranging from 1 to 20 picoseconds, transferring energy through electron-lattice interactions. Subsequently, the lattice of the material experiences heating over a period of 20 to 100 picoseconds, leading to thermal expansion and non-equilibrium conditions. This thermal energy accumulation ultimately drives nonthermal melting and ablation processes, occurring within picoseconds to nanoseconds, where the material is removed through mechanisms such as phase explosion, Coulomb explosion, and spallation, rather than traditional thermal processes like vaporization or melting.

3.1 Spallation

Spallation occurs when the laser energy or fluence is low but still sufficient to induce stress within the material. This ablation mechanism involves the rapid generation of stress waves within the material due to the sudden absorption of laser energy. These stress waves propagate through the material and cause it to fracture and eject material layers from the surface. This dynamic fracture or ejection is called "spallation" and is usually caused by stress and shock waves.

3.2 Phase explosion

When fluence is high, a direct transition of solid material into metastable liquid near its critical state occurs without boiling because of the short heating time. Subsequent bubble nucleation causes a rapid transition of the superheated liquid to a mixture of vapor and liquid droplets being ejected from the bulk material called phase explosion. In numerical solution, it is assumed that heterogeneous nucleation occurs due to phase explosion when the lattice temperature reaches 0.9Tcr (critical temperature) [7].

3.3 Coulomb explosion

In dielectric materials, the bandgap between the conduction band and valence band is typically high, inhibiting the generation of free electrons through processes like multiphoton and avalanche ionization, as discussed in the early section. At a low fluence regime, laser pulses do not have enough energy to excite electrons from the valence band, and at that time, material abated because of the Coulomb explosion. It is an electronic mechanism of material disintegrated from a solid by a charged particle. When a laser beam irradiates on the wide band gap dielectric material, at that time, ions get this energy, and the irradiated surface becomes extremely ionized. The repulsive force between these ionized charge particles exceeds the lattice binding strength, and the atomic bond is broken so that the material is ablated. In metals, coulomb explosion does not occur because the surface charge accumulation is effectively quenched by electronic mobility and suppresses the positive ion explosion [26].

Lewis et al. investigated the dynamics of ultrashort pulses in highly absorbing materials, revealing that ablation occurs through one of three mechanisms: spallation, phase explosion, or fragmentation, depending upon fluence [27]. Zhigilei et al. studied the complex interplay between melting, spallation, and phase explosion during laser pulse ablation, highlighting their simultaneous occurrence and intimate relationship [28]. Patrick et al. developed a phase explosion model that uniquely couples carrier and atom dynamics within a unified Monte Carlo and molecular dynamics scheme [29]. Experimental time-of-flight measurements of wide band gap material have found monoenergetic ion beams emitted from irradiated surfaces, favoring Coulomb explosion over phase explosion [30]. Moreover, Bulgakova et al. explored electrostatic disintegration of surface layers in metals, semiconductors, and dielectrics through the modeling of electronic transport and also concluded that Coulomb explosion is possible in dielectrics [31].

4. Origins and formation mechanisms of LIPSS

The periodic features that appear on the surface of materials after irradiation with one or more laser pulses with fluences approaching the material ablation threshold are known as ripples or LIPSS (laser-induced periodic surface structures). Since Birnbaum discovered LIPSS by irradiating semiconductors with a pulsed ruby laser, LIPSS has gained significant attention due to their ability to modify the surface morphology of materials at the micro-nanoscale and show great promise for surface functionalization applications [32, 33]. LIPSS can be generated on various materials, irrespective of their class, including metals [34], semiconductors [35], dielectrics [36], and polymers [37]. Due to its distinctive characteristics, LIPSS finds applications in various fields, including medicine, optics, tribology, biology, and more [38]. Generally, LIPSS is categorized based on their spatial period and orientation. They are broadly categorized into two main types: low-spatial frequency LIPSS (LSFL) and high-spatial frequency LIPSS (HSFL). Distinguishing between the two, LSFL is characterized by periods larger than half of the laser irradiation wavelength, while HSFL exhibits periods smaller than half of the incident wavelength [39]. Generally, under low fluence conditions, HSFLs are observed, aligned parallel to the polarization of the laser beam. However, as the fluence is increased to a higher regime, LSFL becomes more prominent and exhibits a perpendicular orientation to the laser beam polarization. This transition from HSFL to LSFL occurs as the fluence increases and the number of laser pulses decreases [35, 40]. Moreover, The polarization of the laser beam affects not only the LIPSS orientation but also their periodicity. **Figure** 7 illustrates the formation of LSFL and HSFL on silicon. Notably, LSFL aligns parallel to the electric field, while HSFL exhibits a perpendicular orientation to the electric field.

The periodicity and orientation of LIPSS are affected by irradiation parameters such as laser wavelength [41], angle of incidence [42], polarization [43], the number of pulses [44], and laser fluence [45], as well as by environmental circumstances like pressure, temperature [46], and by material properties such as surface roughness [47]. Experimental results have shown that the LSFL periodicity is proportional to the laser wavelength but can be further tuned with the angle of incidence at which the laser beam hits the target surface. In semiconductors, the periodicity decreases with increasing fluence, a behavior attributed to Surface Plasmon Polaritons (SPPs) [35]. This is in contrast to metals, where an opposite trend is observed—periodicity increases with rising fluence [39, 45]. The larger periodicity of LIPSS at higher fluence in metals is attributed to the induction of surface plasma waves through the parametric decay of laser light [48]. Given the multitude of applications and widespread interest in LIPSS, it has become crucial to understand and confirm the origin of these micro-nanostructures. The first hypothesis on the LIPSS formation was proposed by

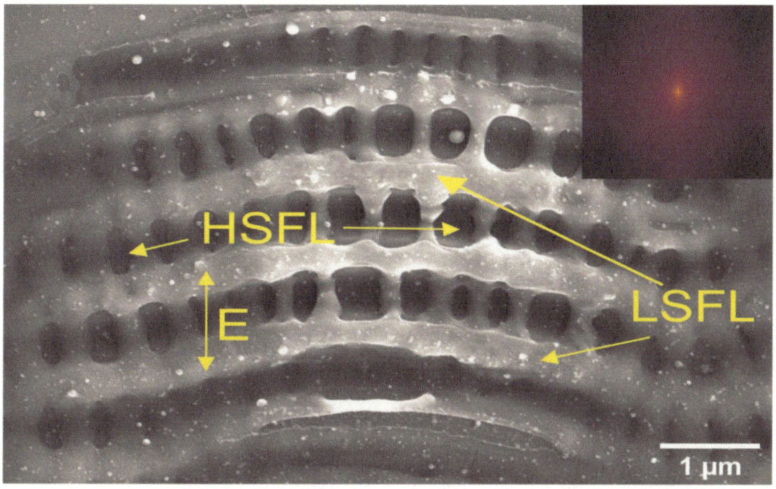

Figure 7.
SEM image showing integrated high-spatial frequency LIPSS (HSFL) and low-spatial frequency LIPSS (LSFL) on silicon at a fluence of 0.90 J/cm^2 and 20 laser pulses. The 2D-FFT analysis is presented in the top right corner of the image.

Birnbaum [32], who attributed their formation to a diffraction effect. Several experimental and theoretical studies have extensively investigated the formation of LIPSS by varying laser and process parameters [44, 49–51]. However, LIPSS formation mechanisms are not yet fully understood. Several mechanisms have been discussed in this section to understand the core principle behind the generation of LIPSS.

4.1 The sipe diffraction model

The first theory based on the formation of LIPSS was given by Sipe, and it has been widely accepted for classical low-spatial frequency LIPSS [50]. Based on this model, LIPSS formed because of the interference between the incident laser beam and Scatter electromagnetic wave (SEW). Sipe theory describes how surface roughness alters the electric field intensity distribution of a plane wave incident on a thin surface with a self-edge region, the height of which is much smaller than the incident laser wavelength (**Figure 8**).

In this theoretical framework, an incident either s or p polarized laser on a rough surface of material with a wave vector K at an angle of incidence θ, projecting a component in the horizontal plane K_i. Sipe's theory predicts the possible wave vectors (k) of LIPSS, related to their period (Λ) through $|k| = \frac{2\pi}{\Lambda}$ [52]. The inhomogeneous energy deposition is directly proportional to $\eta(k, k_i) \cdot |b(k)|$, where $\eta(k, k_i)$ characterizes the efficacy of roughness in contributing to inhomogeneous energy absorption at K, and $|b(k)|$ represents a measure of surface roughness amplitude at k [52, 53].

The efficacy factor η may exhibit sharp peaks at specific k values, allowing evaluation of associated spatial periods Λ. For a surface with uniformly distributed roughness, $|b(k)|$ varies slowly. The Sipe Theory thus provides a comprehensive understanding of how interference patterns, surface roughness, and laser parameters collectively contribute to the intricate process of LIPSS formation. The sipe model, however, has certain limitations due to its high predictive character on processing features. It is an approximation in which the longitudinal component of the electromagnetic field is treated using the variational principle, while the transverse

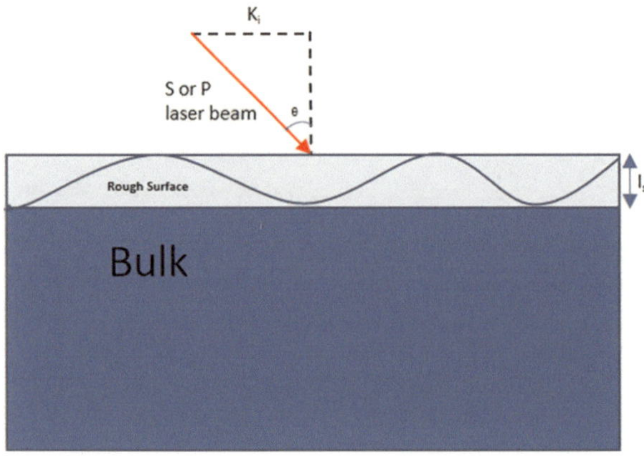

Figure 8.
Schematic representation of laser beam irradiation on a rough surface at an incident angle θ, depicting K_i as the wave-vector component parallel to the surface and Is as the thickness of the rough region. This illustration complements the discussion of laser interaction with rough surfaces in the context of Sipe theory.

component is treated using the perturbation series [52]. Over the years, significant advancements have been made in this theory to understand the formation of LIPSS. This enhanced theoretical framework, often referred to as the Sipe-Drude theory, successfully predicted the features of LIPSS based on irradiation parameters (wavelength, angle of incidence, polarization direction, and laser incident wave vector) as well as surface parameters (dielectric permittivity and surface roughness) [47, 54]. However, the Sipe-Drude theory is well explained for LSFL but an alternative mechanism or theory is needed to understand the HSFL formation.

4.1.1 Surface electromagnetic waves (SEWs) and surface plasmon polaritons (SPPs)

The basis of this model lies in the modulation of the electron on the irradiated surface, particularly through the interaction of the incident laser beam with surface-scattered electromagnetic waves (SEW). Due to the incidence of a laser beam on a metallic, slightly rough surface, scattered fields are generated as a result of surface irregularities. The interference between these scattered fields and the incident or refracted field leads to an inhomogeneous intensity distribution above and below the surface. These inhomogeneous intensity distributions increase the roughness, which contributes to an increase in SEW. As a result, regular patterns start forming, driven by the initial random roughness. In the case of metal-dielectric or metal-air interface, the SEW excite collective oscillations of electrons are known as surface plasmons. These surface plasmons are coupled to the incident light and propagate along the surface, as shown in **Figure 9**. The excitation SSPs involves the periodic modulation of the laser field and subsequent energy deposition in the free electrons [56]. This process heats, melts, or potentially vaporizes the lattice through electron-phonon interaction, subsequently prompting the emergence of LIPSS [56–58].

Based on SPPs theory, the periods are expressed as follows [56]:

$$\Lambda = \frac{\lambda}{\frac{\lambda}{\lambda_{\text{spp}}} \pm \sin(\theta)} \tag{17}$$

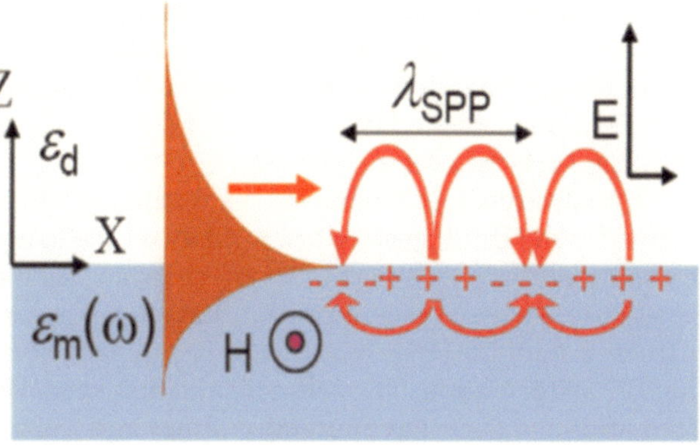

Figure 9.
Schematic representation illustrating the intensity distribution of the electromagnetic field components of a surface Plasmon Polariton (SPPs) supported by a metal-dielectric interface [55].

$$\lambda_{\mathrm{Spp}} = \lambda \times \mathrm{Re}\left(\sqrt{\frac{\varepsilon_{\mathrm{m}} + \varepsilon_{\mathrm{d}}}{\varepsilon_{\mathrm{m}}\varepsilon_{\mathrm{d}}}}\right), \tag{18}$$

where λ_{SPP} is the Surface Plasmon Polariton (SPP) wavelength, θ is the angle of incidence, ε_m is the dielectric constant of the metal, and ε_d is the dielectric constant of the dielectric.

4.2 Self-organization model based on material instability

The foundation of these models rests upon the existing formation theory of ion-beam-induced ripples, which is closely related to laser polarization dependence and ionized kinetic energy distribution [59]. In contrast to previous models, the presented self-organization model places emphasis on the role of hydrodynamics in the formation of LIPSS. **Figure 10** depicts LIPSS formation using the interference model and the

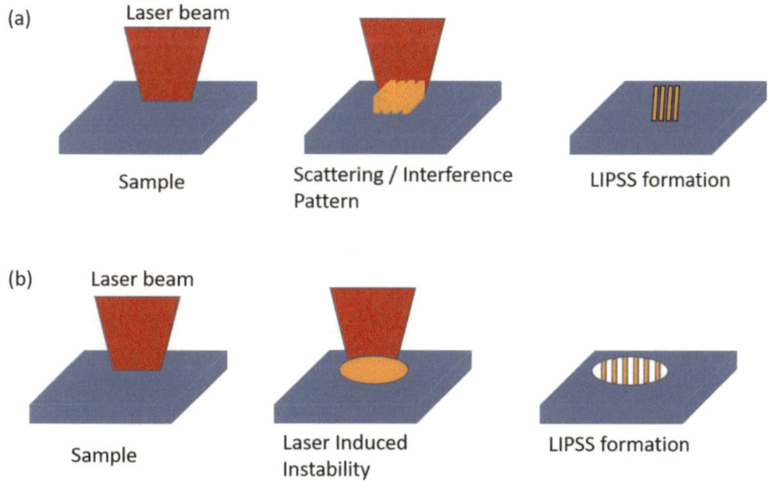

Figure 10.
LIPSS formation occurs via two distinct models: (a) the interference model and (b) the self-organization model [60].

self-organization model. In the interference model, spatially modulated energy deposition creates an "interference" pattern, resulting in periodic pattern formation. On the other hand, in the self-organization model, the surface experiences a high degree of instability induced by a femtosecond laser beam, leading to self-organized periodic pattern formation during surface relaxation.

The key concept in hydrodynamic theories is the formation of ripples through melting flow, triggered by the irradiation of laser pulses on the sample surface. It is crucial to highlight that each proposed instability mechanism, such as Marangoni instability [61], recoil-force-driven instability [62], Rayleigh-Taylor instability [63], and evaporation-driven instability [60], requires an initial perturbation to act as the seed for the ensuing LIPSS formation. The self-organization model has effectively estimated LIPSS formation, considering interaction times and the number of incident pulses. It has also evaluated the relative influence of input energy on the resulting ripple period.

4.3 Finite difference time domain (FTDT) model

The simulation of light propagation, scattering, and diffraction phenomena can be effectively carried out through the finite difference time domain (FDTD) method by solving Maxwell's equation [64]. In this context, Skolski et al. proposed a numerical method based on the FDTD technique to qualitatively investigate LIPSS formation on surfaces, considering factors such as roughness and inhomogeneous energy absorption in the irradiated surface [65]. It has been modified further by employing the inter-pulse feedback effect [66]. Following the initial laser pulse, absorbed electromagnetic energy in the surface layer modifies the material, prompting subsequent material removal. After subsequent laser pulses, the newly morphed surface is utilized for further FDTD simulations, resulting in a new absorbed energy profile and further modification of the surface morphology [66]. The FDTD method simulates the periodicity of LIPSS in semiconductors [66], plasmonic materials [67], and metals [67]. Additionally, it provides the theoretical foundation for the formation of HSFL. However, it does not include incubation effects or thermodynamics of molten materials. There are several other theories on the development of LIPSS, such as matter organization theories [53, 56], including second harmonic generation [52], or transient optical coupling [52].

5. Conclusion

The complex phenomena of ultrashort laser ablation involve a multitude of factors, including the spatial distribution of laser energy, absorption dynamics on the irradiated surface, transient changes in optical response, and the subsequent ablation process. Throughout this chapter, we have thoroughly studied these complexities in the context of metals, dielectrics, and semiconductors. Moreover, we also explore various ablation mechanisms specific to different materials and lasers. Additionally, we provided a description of the LIPSS and their classification in LSFL and HSFL according to the spatial period, along with a discussion on the potential mechanism of LIPSS formation. In summary, this chapter illustrates the basic physical phenomena occurring during the laser-matter interaction.

Author details

Hardik Vaghasiya[1] and Paul-Tiberiu Miclea[1,2]*

1 Martin Luther University Halle-Wittenberg, Germany

2 Fraunhofer Center for Silicon Photovoltaic, Germany

*Address all correspondence to: paul-tiberiu.miclea@csp.fraunhofer.de

References

[1] Auston DH, Eisenthal KB, Hochstrasser RM, Johnson CK, Kaiser W, Laubereau A, et al. Ultrashort Laser Pulses and Applications. Springer; 2013. Available from: https://link.springer.com/book/10.1007/978-3-662-02546-8

[2] Han M, Smith D, Ng SH, Anand V, Katkus T, Juodkazis S. Ultra-short-pulse lasers—Materials—Applications. Engineering Proceedings. 2021;**11**(1):44

[3] Balling P, Schou J. Femtosecond-laser ablation dynamics of dielectrics: Basics and applications for thin films. Reports on Progress in Physics. 2013;**76**(3): 036502

[4] Steen WM, Mazumder J. Laser material processing. Springer Science & Business Media. 2010. Available from: https://link.springer.com/book/10.1007/978-1-84996-062-5

[5] Chen J-K, Latham WP, Beraun JE. Axisymmetric modeling of femtosecond-pulse laser heating on metal films. Numerical Heat Transfer: Part B: Fundamentals. 2002;**42**(1):1-17

[6] Lin Z, Zhigilei LV, Celli V. Electron-phonon coupling and electron heat capacity of metals under conditions of strong electron-phonon nonequilibrium. Physical Review B. 2008;**77**(7):075133

[7] Vaghasiya H, Krause S, Miclea P-T. Fundamental study of a femtosecond laser ablation mechanism in gold and the impact of the ghz repetition rate and number of pulses on ablation volume. Optical Materials Express. 2023;**13**(4): 982-996

[8] Rethfeld B, Ivanov DS, Garcia ME, Anisimov SI. Modelling ultrafast laser ablation. Journal of Physics D: Applied Physics. 2017;**50**(19):193001

[9] Dold C, A. Picosecond Laser Processing of Diamond Cutting Edges. Vol. 688. ETH Zurich; 2013. Available from: https://www.research-collection.ethz.ch/handle/20.500.11850/81504

[10] Byskov-Nielsen J, Savolainen J-M, Christensen MS, Balling P. Ultra-short pulse laser ablation of copper, silver and tungsten: Experimental data and two-temperature model simulations. Applied Physics A. 2011;**103**(2):447-453

[11] Hohlfeld J, Müller JG, Wellershoff S-S, Matthias E. Time-resolved thermoreflectivity of thin gold films and its dependence on film thickness. Applied Physics B. 1997;**64**(3):387-390

[12] Vaghasiya H, Krause S, Miclea P-T. Fundamental study of ablation mechanisms in crystalline silicon and gold by femtosecond laser pulses: Classical approach of two-temperature model. In: Advances in Ultrafast Condensed Phase Physics III. Vol. 12132. SPIE; 2022. pp. 14-25. Available from: https://spiedigitallibrary.org

[13] Vaghasiya H, Krause S, Miclea P-T. Thermal and non-thermal ablation mechanisms in crystalline silicon by femtosecond laser pulses: Classical approach of the carrier density two temperature model. Journal of Physics D: Applied Physics. 2022;**55**(17): 175109

[14] Kaiser A, Rethfeld B, Vicanek M, Simon G. Microscopic processes in dielectrics under irradiation by subpicosecond laser pulses. Physical Review B. 2000;**61**(17):11437

[15] Keldysh LV et al. Ionization in the field of a strong electromagnetic wave.

Soviet Physics JETP. 1965;**20**(5):
1307-1314

[16] Bulgakova NM, Zhukov VP. Continuum models of ultrashort laser–matter interaction in application to wide-bandgap dielectrics. In: Lasers in Materials Science. Springer; 2014. pp. 101-124. Available from: https://link.springer.com/chapter/10.1007/978-3-319-02898-9_5

[17] Feng T, Chen G, Han H, Qiao J. Femtosecond-laser-ablation dynamics in silicon revealed by transient reflectivity change. Micromachines. 2021;**13**(1):14

[18] Rämer A, Haahr-Lillevang L, Rethfeld B, Balling P. Modeling the transient optical parameters in laser-excited band gap materials. Optical Engineering. 2017;**56**(1):011015-011015

[19] Chen JK, Beraun JE. Numerical study of ultrashort laser pulse interactions with metal films. Numerical Heat Transfer: Part A: Applications. 2001;**40**(1):1-20

[20] Jiang L, Tsai H-L. Improved two-temperature model and its application in ultrashort laser heating of metal films. Journal of Heat Transfer. 2005;**127**: 1167-1173

[21] Qiu TQ, Tien CL. Short-pulse laser heating on metals. International Journal of Heat and Mass Transfer. 1992;**35**(3): 719-726

[22] Sobolev SL. Nonlocal two-temperature model: Application to heat transport in metals irradiated by ultrashort laser pulses. International Journal of Heat and Mass Transfer. 2016; **94**:138-144

[23] Shugaev MV, Gnilitskyi I, Bulgakova NM, Zhigilei LV. Mechanism of single-pulse ablative generation of laser-induced periodic surface structures. Physical Review B. 2017; **96**(20):205429

[24] Wang X, Ye X, Yao H, Wei P, Yin F, Cong J, et al. Simulation of femtosecond laser ablation and spallation of titanium film based on two-temperature model and molecular dynamics. Journal of Laser Applications. 2021;**33**(1):012047

[25] von der Linde D, Sokolowski-Tinten K. Physical mechanisms of short-pulse laser ablation. In: ICONO'98: Fundamental Aspects of Laser-Matter Interaction and New Nonlinear Optical Materials and Physics of Low-Dimensional Structures. Vol. 3734. SPIE; 1999. pp. 2-9. Available from: https://spiedigitallibrary.org

[26] Samad RE, Machado LM, Vieira ND Jr, De Rossi W. Ultrashort laser pulses machining. In: Laser Pulses-Theory, Technology, and Applications. London, UK: IntechOpen; 2012. pp. 143-174

[27] Lewis LJ, Perez D. Laser ablation with short and ultrashort laser pulses: Basic mechanisms from molecular-dynamics simulations. Applied Surface Science. 2009;**255**(10):5101-5106

[28] Zhigilei LV, Lin Z, Ivanov DS. Atomistic modeling of short pulse laser ablation of metals: Connections between melting, spallation, and phase explosion. The Journal of Physical Chemistry C. 2009;**113**(27):11892-11906

[29] Lorazo P, Lewis LJ, Meunier M. Short-pulse laser ablation of solids: From phase explosion to fragmentation. Physical Review Letters. 2003;**91**(22):225502

[30] Henyk M, Wolfframm D, Reif J. Ultra short laser pulse induced charged particle emission from wide bandgap crystals. Applied Surface Science. 2000; **168**(1–4):263-266

[31] Bulgakova NM, Stoian R, Rosenfeld A, Hertel IV, Campbell EEB. Electronic transport and consequences for material removal in ultrafast pulsed laser ablation of materials. Physical Review B. 2004;**69**(5):054102

[32] Birnbaum M. Semiconductor surface damage produced by ruby lasers. Journal of Applied Physics. 1965;**36**(11):3688-3689

[33] Bonse J, Kirner SV, Höhm S, Epperlein N, Spaltmann D, Rosenfeld A, et al. Applications of laser-induced periodic surface structures (lipss). In: Laser-Based Micro-and Nanoprocessing XI. Vol. 10092. SPIE; 2017. pp. 114-122. Available from: https://journals.aps.org/prb/abstract/10.1103/PhysRevB.69.054102

[34] Zhang D, Liu R, Li Z. Irregular lipss produced on metals by single linearly polarized femtosecond laser. International Journal of Extreme Manufacturing. 2021;**4**(1):015102

[35] Vaghasiya H, Miclea P-T. Investigating laser-induced periodic surface structures (lipss) formation in silicon and their impact on surface-enhanced raman spectroscopy (Sers). Optics. 2023;**4**(4):538-550

[36] Fang Z, Shao J, et al. Femtosecond laser-induced periodic surface structure on fused silica surface. Optik. 2016; **127**(3):1171-1175

[37] Mezera M, van Drongelen M, Römer GRBE. Laser-induced periodic surface structures (lipss) on polymers processed with picosecond laser pulses. Journal of Laser Micro Nanoengineering. 2018;**13**(2):105-116

[38] Gräf S. Formation of laser-induced periodic surface structures on different materials: Fundamentals, properties and applications. Advanced Optical Technologies. 2020;**9**(1–2): 11-39

[39] Bonse J. Quo vadis lipss?—Recent and future trends on laser-induced periodic surface structures. Nanomaterials. 1950;**10**(10):2020

[40] Shi X, Xuefeng X. Laser fluence dependence of ripple formation on fused silica by femtosecond laser irradiation. Applied Physics A. 2019;**125**:1-8

[41] Li G, Li J, Yanlei H, Zhang C, Li X, Chu J, et al. Femtosecond laser color marking stainless steel surface with different wavelengths. Applied Physics A. 2015;**118**:1189-1196

[42] Nürnberger P, Reinhardt H, Kim HC, Yang F, Peppler K, Janek J, et al. Influence of substrate microcrystallinity on the orientation of laser-induced periodic surface structures. Journal of Applied Physics. 2015;**118**(13):134306

[43] Murphy R, D, Torralva B, Adams DP, Yalisove SM. Polarization dependent formation of femtosecond laser-induced periodic surface structures near stepped features. Applied Physics Letters. 2014;**104**(23)

[44] Bonse J, Höhm S, Rosenfeld A, Krüger J. Sub-100-nm laser-induced periodic surface structures upon irradiation of titanium by ti: Sapphire femtosecond laser pulses in air. Applied Physics A. 2013;**110**:547-551

[45] Okamuro K, Hashida M, Miyasaka Y, Ikuta Y, Tokita S, Sakabe S. Laser fluence dependence of periodic grating structures formed on metal surfaces under femtosecond laser pulse irradiation. Physical Review B. 2010; **82**(16):165417

[46] Gräf S, Kunz C, Engel S, Derrien TJ-Y, Müller FA. Femtosecond laser-induced periodic surface structures on fused silica: The impact of the initial substrate temperature. Materials. 2018; **11**(8):1340

[47] Fuentes-Edfuf Y, Sanchez-Gil JA, Florian C, Giannini V, Solis J, Siegel J. Surface plasmon polaritons on rough metal surfaces: Role in the formation of laser-induced periodic surface structures. ACS Omega. 2019;**4**(4): 6939-6946

[48] Sakabe S, Hashida M, Tokita S, Namba S, Okamuro K. Mechanism for self-formation of periodic grating structures on a metal surface by a femtosecond laser pulse. Physical Review B. 2009;**79**(3):033409

[49] Bonse J, Munz M, Sturm H. Structure formation on the surface of indium phosphide irradiated by femtosecond laser pulses. Journal of Applied Physics. 2005;**97**(1):013538

[50] Sipe JE, Young JF, Preston JS, Van Driel HM. Laser-induced periodic surface structure. I. Theory. Physical Review B. 1983;**27**(2):1141

[51] Wang J, Guo C. Ultrafast dynamics of femtosecond laser-induced periodic surface pattern formation on metals. Applied Physics Letters. 2005;**87**(25): 230013

[52] Li C, Stoian R, Cheng G-H, et al. Laser-induced periodic surface structures with ultrashort laser pulse. Chinese Optics. 2018;**11**(1):1-17

[53] Bonse J, Gräf S. Maxwell meets marangoni—A review of theories on laser-induced periodic surface structures. Laser & Photonics Reviews. 2020;**14**(10):2000215

[54] Dufft D, Rosenfeld A, Das SK, Grunwald R, Bonse J. Femtosecond laser-induced periodic surface structures revisited: A comparative study on zno. Journal of Applied Physics. 2009;**105**(3): 034908

[55] Han Z, Bozhevolnyi SI. Waveguiding with surface plasmon polaritons. In: Handbook of Surface Science. Vol. 4. Elsevier; 2014. pp. 137-187. Available from: https://www.sciencedirect.com/science/article/abs/pii/B9780444595263000057

[56] Zhang Y, Jiang Q, Long M, Han R, Cao K, Zhang S, et al. Femtosecond laser-induced periodic structures: Mechanisms, techniques, and applications. Opto-Electronic Science. 2022;**1**(6):220005

[57] Bonse J, Rosenfeld A, Krüger J. On the role of surface plasmon polaritons in the formation of laser-induced periodic surface structures upon irradiation of silicon by femtosecond-laser pulses. Journal of Applied Physics. 2009; **106**(10):104910

[58] Huang M, Zhao F, Cheng Y, Ningsheng X, Zhizhan X. Origin of laser-induced near-subwavelength ripples: Interference between surface plasmons and incident laser. ACS Nano. 2009; **3**(12):4062-4070

[59] Mark Bradley R, Harper JME. Theory of ripple topography induced by ion bombardment. Journal of Vacuum Science & Technology A: Vacuum, Surfaces, and Films. 1988;**6**(4): 2390-2395

[60] Varlamova O, Reif J, Varlamov S, Bestehorn M. Self-organized surface patterns originating from laser-induced instability. In: Progress in Nonlinear Nano-Optics. Springer; 2015. pp. 3-29. Available from: https://link.springer.

com/chapter/10.1007/978-3-319-12217-5_1

[61] Tsibidis GD, Barberoglou M, Loukakos PA, Stratakis E, Fotakis C. Dynamics of ripple formation on silicon surfaces by ultrashort laser pulses in subablation conditions. Physical Review B. 2012;**86**(11):115316

[62] Gurevich EL. Mechanisms of femtosecond lipss formation induced by periodic surface temperature modulation. Applied Surface Science. 2016;**374**:56-60

[63] Birkhoff G. Hydrodynamics. Vol. 2234. Princeton University Press; 2015. Available from: https://www.degruyter.com/document/doi/10.1515/9781400877775/html?lang=en

[64] Yee K. Numerical solution of initial boundary value problems involving maxwell's equations in isotropic media. IEEE Transactions on Antennas and Propagation. 1966;**14**(3):302-307

[65] Skolski JZP, Römer GRBE, Vincenc Obona J, Ocelik V, De Hosson JTM, et al. Laser-induced periodic surface structures: Fingerprints of light localization. Physical Review B. 2012; **85**(7):075320

[66] Skolski JZP, Römer GRBE, Vincenc Obona J, et al. Modeling laser-induced periodic surface structures: Finite-difference time-domain feedback simulations. Journal of Applied Physics. 2014;**115**(10):1325-1333

[67] Zhang H, Colombier J-P, Li C, Faure N, Cheng G, Stoian R. Coherence in ultrafast laser-induced periodic surface structures. Physical Review B. 2015;**92**(17):174109

Chapter 3

Epitaxial Ferroelectric Thin Films: Potential for New Applications

Cristina Chirila, Andra G. Boni, Lucian D. Filip, Mihaela Botea, Dana Popescu, Viorica Stancu, Lucian Trupina, Luminita Hrib, Raluca Negrea Ioana Pintilie and Lucian Pintilie

Abstract

This chapter provides an overview of the versatile applications and properties of epitaxial ferroelectric materials obtained using the pulsed laser deposition technique. These materials can play a significant role in various electronic and sensing applications or energy harvesting. Materials that are ferroelectric and have a perovskite structure (ABO3 type) show spontaneous polarization that can be changed by an electric field, temperature, mechanical stress, or light. Here, we present results obtained on epitaxial ferroelectric thin films with different compositions, lead-based or lead-free, and the correlation with structural quality of the layers and with different electrostatic conditions induced either by the substrate or by the different dopants. Our studies revealed that the utilization of pulsed laser beam deposition technique is suitable for obtaining ultrathin films depositions with thicknesses measuring less than 5 nm. These results allowed us to reveal the impacts caused by polarization orientation on the band structure or the presence of self-doping phenomena. We also found that the conduction type can be modified by introducing 1% Fe and Nb on $PbZrTiO_3$ (PZT) epitaxial layers. In the last part of this chapter, we report on obtaining of a lead-free epitaxial thin film and its properties in the energy storage field.

Keywords: pulsed laser deposition, epitaxy, thin films, lead-based, lead-free, supercapacitors

1. Introduction

The complexity of electronic devices has increased due to the development of advanced technologies such as 5G, the Internet of Things (IoT), and artificial intelligence. However, due to consumer preferences for portable gadgets and technological advancements, there is a trend toward smaller and more compact electronic devices. This trend requires materials that can maintain or enhance performance while occupying less space. At the same time, as more and more people become concerned about environmental impact and power consumption, there is a demand for materials that can aid in the creation of more efficient electrical equipment. Such materials are the ones that possess a spontaneous electric polarization P, which is reversible and stable over time under external electric fields greater than the coercive field (i.e., ferroelectrics). The paraelectric to ferroelectric transition is known to entail a structural change from high to low symmetry, which causes the atoms to be off-centered and displaced from their symmetric orientations [1]. Ferroelectric thin films are used in many electronic and sensing applications, such as ferroelectric field effect transistors, pyroelectric infrared detectors, and supercapacitors. Epitaxial thin films offer advantages in terms of controlled growth and strain engineering, which increases their ability to be tailored for specific applications. When targeting an application that employs functional elements, it is crucial to examine the mechanisms, both intrinsic and extrinsic, that play a role in stabilizing a clearly defined ferroelectric state. It is also vital to separate and analyze how these mechanisms affect electrical properties. Currently, magnetron sputtering and pulsed laser deposition (PLD) provide good results with respect to high epitaxial quality [2]. But there is still more to be done: to make the targets more chemically pure, to keep the deposition chamber free of contaminants, precisely control the reactive gas level, and to make sure that the substrates are chemically and thermally treated prior to deposition, with uniform terraces.

The choice of substrate is crucial in PLD to obtain high-quality epitaxial thin films with a well-defined crystal orientation. Monocrystalline substrates, such as $SrTiO_3$ (STO), $LaAlO_3$ (LAO), and MgO, are commonly used for epitaxial thin films due to their similar lattice structures and low misfit strain [3, 4]. The choice of bottom electrode material has a substantial impact on the ferroelectric properties and total losses, following the influence of the substrate. The bottom electrode can be either a basic metal or a conductive oxide. The use of conducting oxide materials as bottom electrodes enables the creation of a ferroelectric layer with minimum imperfections and strain mismatch [3, 5].

Here, $SrRuO_3$ (SRO) and $LaSrMnO_3$ (LSMO) conductive oxides have been used like bottom electrodes for all structures. Even though lead-based ferroelectric materials are widely studied and used in solid-state technology, research into their fundamental properties, such as the conduction type in ferroelectric films or their intrinsic band structure, is still up for debate. Here, we have investigated these fundamental characteristics in $PbZrTiO_3$ (PZT) epitaxial layers with different thicknesses, and the electrical characterizations were performed on capacitor-like geometry. For this, top SRO/Au electrodes were deposited by PLD and magnetron sputtering, with a shadow mask, defining ferroelectric capacitors of 100 μm^2 area. We found also that one way to tailor the PZT properties for certain purposes is through doping.

Doping effects must be studied on materials of excellent crystalline quality, ideally approaching single crystal, to avoid masking effects caused by structural defects, such as grain boundaries. PZT single crystals are challenging to synthesize due to PbO volatility during growth. However, high-quality epitaxial layers can be grown using methods, such as pulsed laser deposition (PLD) on suitable single-crystal substrates, such as $SrTiO_3$ (STO). This advancement enables the study of dopants' effects on electrical properties in epitaxial PZT layers at doping levels around or below 1% atomic. The following results show the impact of 1% Fe and Nb doping on the electrical properties of epitaxial PZT films.

Simultaneously, due to concerns about lead-based materials, particularly their environmental impact, we investigated alternate materials. One such material is bismuth-based perovskite compounds, which have been found to exhibit properties useful on energy storage, as well as being nontoxic and environmentally friendly.

$Bi_{0.5}Na_{0.5}TiO_3$ (BNT) is considered one of the candidates as lead-free material for electrical energy storage due to its high dielectric constant and excellent ferroelectric properties. Until now, a large number of publications on the energy storage properties of lead-free ferroelectrics are related to ceramics or polycrystalline thin films [6–9]. Here, the lead-free ferroelectric epitaxial thin films have been developed and researched.

2. Pulsed laser deposition

Due to its flexibility, PLD method is now widely used for the deposition of oxide materials, especially for research, and particularly of multicomponent oxides. This technique offers very good translation of stoichiometry from the target to the deposited film and proved to be already scalable for deposition on large area wafers in the future [10, 11]. PLD process also depends on the type of laser used. It must consider the absorption coefficient and reflectivity of the materials that will be deposited. Because these properties vary with wavelength, the laser must be able to operate in the wavelength range where the target material has low reflectivity and high absorption coefficient for efficient deposition. Excimer lasers (XeCl, KrF, and ArF) are widely utilized for the deposition of complicated oxide films because, at the wavelengths used by these lasers, these materials have a high absorption coefficient and low reflectance [12]. The process of PLD involves several key steps. First, a high-intensity laser pulse is directed onto a target material composed of the desired oxide. This laser pulse vaporizes or ablates the target, creating a plasma plume consisting of atoms, ions, and clusters of the target material [13]. The substrate, typically positioned parallel to the target is then exposed to this plasma, allowing the deposition of the ablated material onto its surface, **Figure 1**.

The growth dynamics of thin films produced by PLD are highly dependent on various parameters such as target composition, substrate temperature, laser fluence and frequency, the spot size, and the background pressure [14]. Multiple targets are contained in the target holder, four in our system, enabling the deposition of multilayer thin films. In order to prevent the formation of droplets on the surface of the films during the laser deposition process, it is preferable to use a target with high density, purity, and a reduced laser pulse frequency.

Deposition often involves constant translation and rotation of the target support. The substrate heating block, on our system, has the capability to reach temperatures as high as 1000°C. The experimental results showed that there is a critical substrate

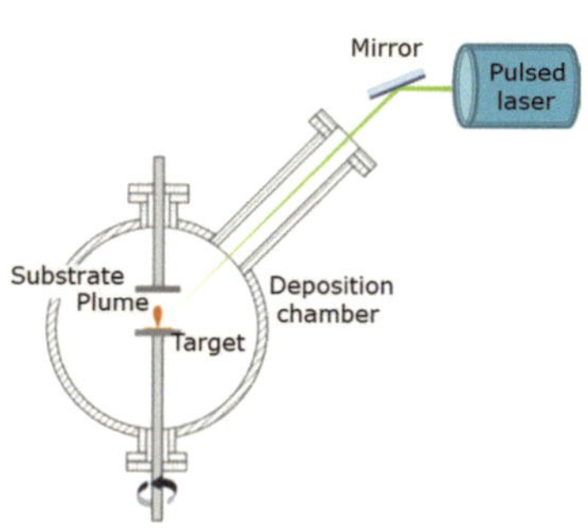

Figure 1.
Schematic representation of pulsed laser deposition with parallel setup.

temperature (Tc) below which the film structure is not completely monocrystalline and the film composition deviates significantly from the stoichiometric one [15, 16]. The optimal frequency of laser pulses for film preparation depends on the ambient gas pressure and the substrate-target distance [17, 18]. The interaction between the plasma and the substrate has a major impact for deposition. A balance between the target heating speed and substrate temperature is required to achieve the desired film composition and properties. Short-duration pulses (10^{-3} s) and energy densities of 10^6–10^8 W/cm^2 allow for the achievement of heating speeds of 10^{10} K/s [13, 14] at the target surface. The heat loss of the components arriving on the substrate is determined by the substrate temperature. The rapid cooling of the compound causes it to crystallize at a slower rate. Due to the extremely short distances between atoms on the surface of the substrate, only a small number of them are able to form crystals.

The temperature of the layer, thus, formed increases when the next vapor pulse is deposited. The impact of the impinging flux of ablated particles on this temperature becomes more significant at higher deposition frequencies and should be considered, particularly when considering the formation of materials with low thermal conductivity using PLD [19].

If the substrate temperature is sufficiently high, the material's velocity at the substrate surface will decrease, hence facilitating the optimal conditions for complete crystallization, nucleation, and coalescence of the atoms [15, 20]. To ensure proper stoichiometry, the rate of arrival of the constituents of one complex compound must be carefully managed, and this rate is also being affected by the gaseous environment. The gas in the deposition chamber affects the crystalline microstructure and orientation by its chemical activity, which is naturally proportional to the pressure [15]. Complex oxide films are typically prepared in oxygen atmosphere at a pressure of several hundred millitorr. The use of high oxygen pressure is employed for the deposition of complex oxide films, but above a certain point, it can have the opposite effect of what you would expect by slowing down the deposition rate due to oxygen's high dissociation energy barrier, which makes it chemically reactive [4, 16, 21, 22].

3. Termination control for epitaxial thin films

Achieving epitaxial growth (where the crystal structure of the film aligns with that of the substrate) can be challenging if there are significant differences in lattice parameters [23, 24]. The lattice constant (spacing between atoms) of the thin film may not perfectly match that of the substrate. This can result in strain and defects in the film, affecting its structural properties. Prior to epitaxial growth, the substrate surface needs to be carefully prepared to achieve the desired termination. Substrate termination can impact the nucleation process, affecting the initial stages of thin film growth [25–27]. Controlling nucleation is crucial for achieving a high-quality and uniform epitaxial layer. Techniques, such as chemical etching, annealing, or other surface treatments, are employed to modify the crystal surface and remove contaminants. The examined structures in this chapter were developed using single crystalline STO (001) and Nb: STO (001) substrates supplied by CrysTec GmbH. These substrates had a miscut angle ranging from 0.05° to 0.5°. Prior to achieving higher quality thin films on single crystalline STO and Nb:STO substrates, preliminary substrates preparation were performed. These transformations involve transforming the optically polished surface into a surface with steps and terraces that are highly organized at the atomic level. To achieve this result, the STO substrates required etching in an

NH4-HF solution for a duration of 15 seconds to eliminate any Sr. residues. Following that, the substrates were placed under a process of thermal treatment at a high temperature of around 1000°C for a period of about 4 hours. By following this procedure, a surface consisting completely of TiO_2 is achieved, with approximately the same height steps (about 0.4 nm per unit cell), having parallel position to each other, are selected. The miscut angle value has an important effect on the surface preparation process. While both chemical and thermal annealing follow the same procedure way, we found that different substrates miscut produce different terraces; especially those substrates having a smaller miscut, we noticed that adjusting the annealing period improved the terrace quality (**Figure 2**). Other groups also reported similar behavior on monocrystalline substrates, which have been found to be correlated with the miscut angle value [28, 29]. However, it was challenging to determine the exact times terrace obtaining.

Mismatches between oxidic thin films and monocrystalline substrates can be minimized by introducing a buffer layer with intermediate properties between the thin film and substrate. Buffer layers also provide an approach for achieving preferential polarization orientation; they can contain free charges induced by point defects that form during the deposition process. The presence of these defects acts for compensating the depolarization field and maintain the upward polarization direction [30, 31]. In this chapter, conductive oxides, such as LSMO and SRO, were utilized as buffer materials and bottom electrodes in order to form structures similar to capacitors. The electric and magnetic characteristics of LSMO are dependent by several parameters, including La, Sr content, deposition temperature, thickness, and the crystalline structure of the layer [32–34]. The choice of using $La_{70}Sr_{30}MnO_3$ as the bottom electrode was based on our previous study of the resistivity properties, particularly its dependence on thickness [33].

3.1 Conditions for buffer layer preparation

A KrF laser (λ = 248 nm) was utilized to ablate Praxair's commercial SRO and LSMO targets for buffer layer deposition. The repetition rate for LSMO and SRO was 1 and 5 Hz, respectively, with a laser fluence of 2 J/cm^2. During deposition, the substrate temperature was set to 700°C, with an oxygen atmosphere of 0.133 mbar for SRO and 0.27 mbar for LSMO. The target and substrates were kept at a distance of 60 mm for both types of buffer layers. Further, on these structures, lead-based or lead-free layers with different thicknesses were deposited.

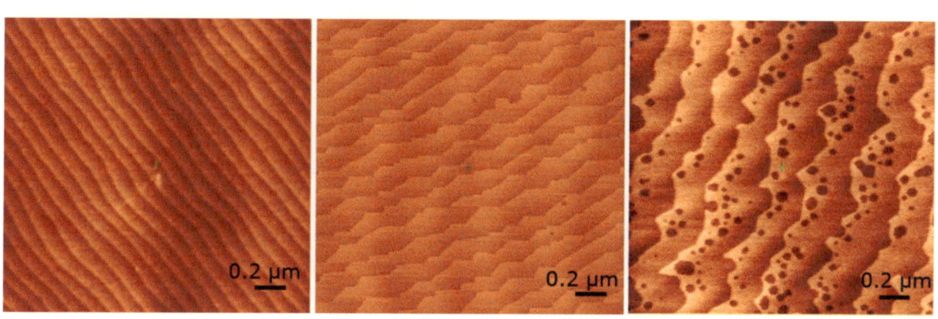

Figure 2.
Atomic force microscopy images from identically cleaned substrates with varying miscut angles.

3.2 Lead-based ferroelectric films: ultrathin films

In this section, we describe our investigations related to the preparation of 5 nm thick PZT ultrathin films by using a commercial target from Praxair with a Zr doping x = 0.2. The films were grown on different ferroelectric states, and the 3D band structure was studied using SX-ARPES [35]. By discerning between the effects caused by ferroelectricity and those induced by the substrate on the electronic band structure, we are able to elucidate basic questions concerning the fundamental electronic properties of oxide interfaces and provide further avenues for enhancing the performance of multiferroic systems. The PZT films were grown on LSMO-buffered TiO_2-terminated STO substrates (001) oriented. In this case, the ferroelectric polarization points toward the interface (from herein named DW sample and denoted as P-). Samples grown on TiO_2-terminated, Nb:STO substrate feature an upwards polarization state (P+) and will be designated as the UP sample, with an out-of-plane FE state, mostly single domain. The growth conditions for PZT films were similar to those detailed in our previous work [35].

3.3 Structural characterization

The symmetric 2θ-ω XRD scans, which are displayed in **Figure 3a**, performed at room temperature on a Rigaku-SmartLab diffractometer in high-resolution setting (X-ray mirror and two bounce Ge (220) monochromator) demonstrate the good crystallographic resemblance of both PZT films. However, DW has somewhat wider peaks than UP, possibly because DW has larger internal microstrains [36]. The PZT films in both DW and UP are fully strained at the in-plane lattice constant of the cubic STO and Nb:STO substrates, as indicated by the RSM near the −103 node of STO (**Figure 3b**) and Nb:STO, **Figure 3c**, performed on a Bruker D8 Advance diffractometer (Bruker AXS GmbH Germany) with a copper anode X-ray tube in medium resolution parallel beam setting. PZT exhibits elongated growth along the c-axis in both samples, with the ratio of the in-plane to out-of-plane lattice parameters c/a = 1.073 for UP and 1.081 for DW. The significantly off-centering of the Pb and Ti/Zr cations

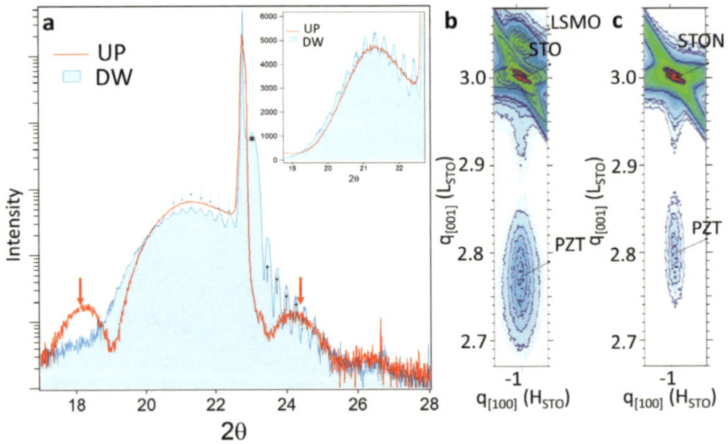

Figure 3.
(a) X-ray diffraction. 2θ-ω scan for PZT grown in UP and DW samples; in the inset, the zoomed region of the 001 peak of PZT in linear scale of intensity is given for both samples; (b), (c) reciprocal space mapping (RSM) scans of DW and UP samples are presented in and, respectively [35].

in the unit cell ensures the stabilization of the FE character of the thin films. The local PFM measurements (**Figure 4**) on the two samples determined the different out-of-plane FE polarization, pointing downward or outward. The outcome confirms that PZT in the DW sample has indeed the FE polarization orientated inward (P−), whereas PZT in the UP sample has the FE state oriented away from the surface (P+).

3.4 k-resolved valence band structure

By navigating in the out-of-plane k_z direction, we first examine the k-space topology of PZT in the UP sample with tetragonal (TG) unit cell. The photon energy $h\nu$ in **Figure 3a** is adjusted at constant $k_{||}$ to range from 350 to 520 eV. The resulting iso-energy map, shown in **Figure 5a**, in the XΓZ plane of the bulk Brillouin zone, locates the valence band maximum (VBM) at a binding energy (BE) of 2.2 eV with respect to the Fermi level in the X point. According to the iso-energy surface obtained from DFT calculations for the tetragonal PZT unit cell, **Figure 5c**, the square-like symmetry of the PZT in the *ab* plane accounts for the ($k_x - k_y$) iso-energy map observed on the UP sample in the XΓM plane at the VBM using hv = 520 eV (**Figure 5b**).

The DFT-predicted pattern centered in the Z point for the tetragonal unit cell, **Figure 5e**, is also visible on the iso-energy map taken in the ZAR plane 0.5 eV below VBM with $h\nu$ = 465 eV (**Figure 5d**). Nevertheless, for the DW sample, the iso-energy estimated in the tetragonal cell as depicted in **Figure 6a** is inconsistent with the iso-energy map obtained in the ZAR plane 0.5 eV below VB. It reveals four elliptical-shaped features centered in the A points of the k-space in addition to the predicted signature of the hole-like band in the Z point. Their appearance is consistent with a rhombohedral (RH) reconstruction of the PZT unit cell, as shown by the 3D iso-E surface of PZT, **Figure 6b**. The primary axis of the pocket surrounding the A point, as indicated by the computed iso-E map, is derived from the projection of the RH cell's WXW direction on the TG-cell's (001) direction. Consequently, projecting the RH cell's UXU direction onto the (001) plane of the TG-cell BZ yields the minor axis of the A-centered ellipse.

3.5 Polarization-dependent band alignment mechanism

We have chosen the two distinct substrates because of their correspondingly larger and smaller work functions (W_f). The migration of negative (positive) charges

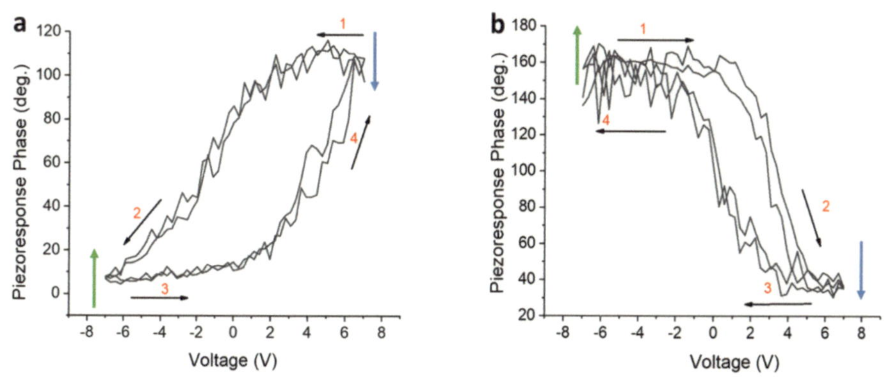

Figure 4.
(a) Piezoresponse phase loops versus applied DC voltage, measured with an AC driving voltage Vac = 1.5 V for DW and (b) UP samples [35].

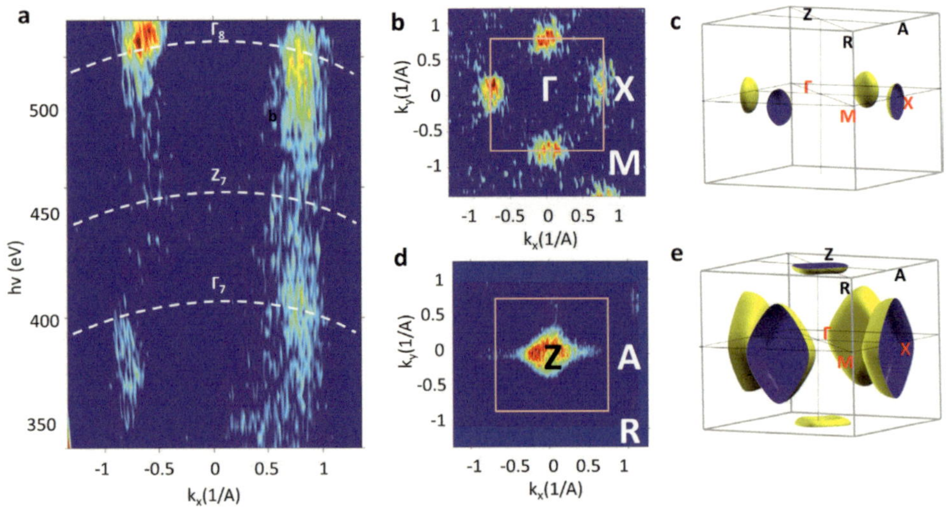

Figure 5.
(a) Out-of-plane iso-energy (iso-E) maps of PZT recorded by varying hv between 350 and 520 eV while keeping the $k_{||}$ in the XΓZ plane of the bulk Brillouin zone; (b) $(k_x - k_y)$ iso-E in the ΓMX plane at the VB maximum—VBM—and (d) in the ZAR plane at 0.5 eV below the VBM. The calculated iso-E surface at (c) VBM and (e) 0.5 eV below VMB [35].

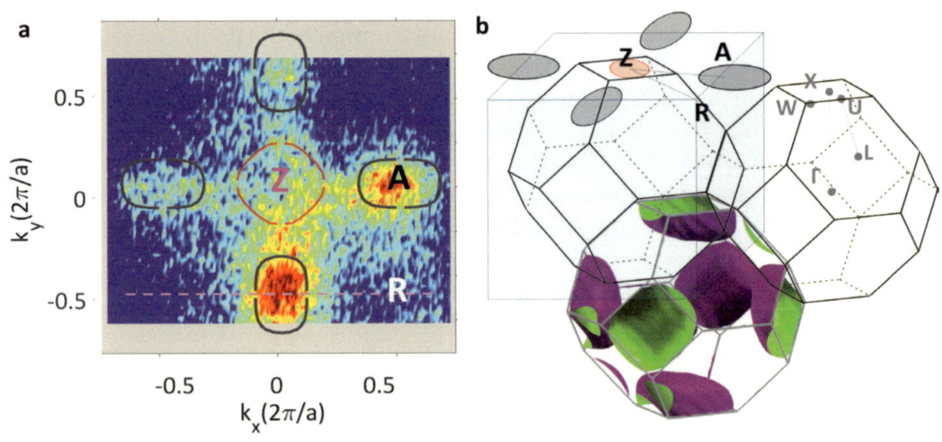

Figure 6.
(a) Signature of RH reconstruction. In-plane (k_x, k_y) iso-E maps of PZT recorded with hv = 465 eV at 0.5 eV below the VBM in the ZAR plane of the TG cell. The red line indicates the calculated iso-contour at the corresponding energy assuming TG geometry of the unit cell. Gray contours represent the additional signature of the RH-distorted unit cell for DW, and such a signature is absent for the UP sample; (b) the signature of the RH distortion is rendered into the TG unit cell, with the iso-energy surfaces calculated at the same energy, 0.5 eV below VBM, as the experimental maps [35].

from PZT at the bottom interface is driven by the material-dependent band bending, $\Delta\phi^{DW(UP)}$, defined by the W_f-induced band lineup at the interface with the substrate. We anticipated the stability of distinct, opposing out-of-plane P−(P+) FE states of PZT in conjunction with the accessible positive (in LSMO) and negative (in Nb:STO) charges in the substrate.

The contaminated layer is identified as the cause of improper screening for the depolarizing field at the PZT surface. Based on the examination of the core levels recorded on both UP and DW samples, we will now clarify the mechanism of band alignment. Thus, by opposing the FE-induced field, the residual uncompensated field tends to lower the band bending potential at the surface. Thin films use internal or external mechanisms to make up for the depolarizing field in order to maintain their ferroelectric state. The intrinsic one entails the migration of charge carriers that are already present in the film, producing positive and negative charge sheets at the layer's opposing surfaces.

These charge carriers are produced when the ideal stoichiometry spontaneously changes during growth, and cations or oxygen vacancies are formed [31]. In the external material, such as the metallic electrode, or within the contaminated layer at the surface with the air to screen the DF, the extrinsic one entails opposite charge accumulation or depletion with respect to the fixed polarization charges from the FE material. As the electric field outside the ferroelectric is strictly zero, the charge modulation effect produced into the joining material is greater than a pure electrostatic picture and should not affect the joining material's electronic structure. However, it is dependent upon two elements: (i) on the specific band alignment at the interfaces between the FE and the metal or (ii) the FE and the contamination layer, which takes into account both the FE state and the work function difference between the FE and the metallic contact or surface contamination layer [37–40].

The regions with positive/negative charge buildup at the ferroelectric's extremities are defined by the band alignment and the potential profile that results across the ferroelectric. It regulates the application of bias, determining whether and how the necessary electrons and/or holes are injected across the interface and how they collect at the ferroelectric's surface and interfaces *via* diffusing through the material [37–39, 41, 42]. **Figure 7a** shows the mechanism of band alignment based on the FE state of the PZT samples at the surface and at their interface with the substrate. It is derived by comparing the more bulk-sensitive results of the core levels, which extend the probing region down to the substrate interface, with the surface-sensitive ARPES data. This shows the variation of the potential inside the FE film going from the surface toward the interface with the substrate.

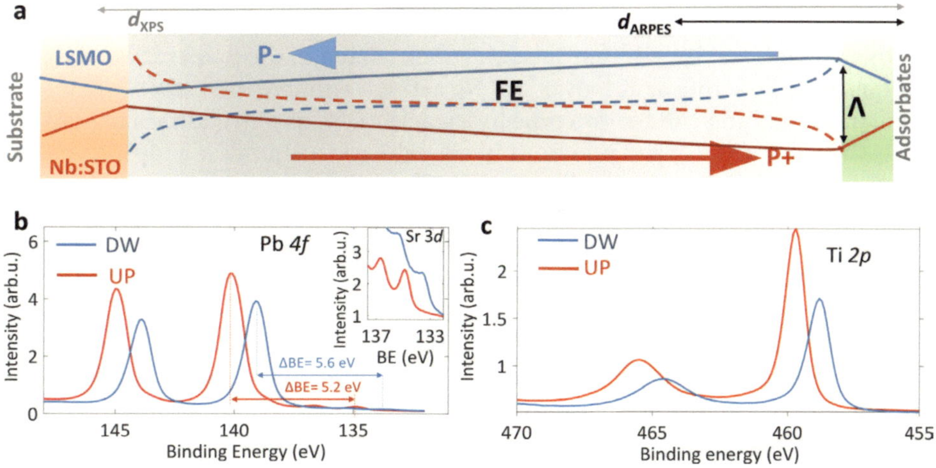

Figure 7.
(a) Band bending mechanism from the core levels. Band bending at the PZT surface as a function of the ferroelectric state; (b, c) sketch of the band bending potential along the UP and DW samples Pb 4f and Ti 2p core level spectra of DW (blue) and UP (red). Samples recorded at 12 K and after 20 min of continuous exposure to the X-ray beam [35].

The potential profile when opposing charges build up in the ferroelectric film near the interfaces for depolarizing filed screening is shown by the dashed line. The potentials V(z) inside our ferroelectric films, for both UP and DW samples, are qualitatively traced with full lines based on the experimental results. The ARPES observations reveal a 1.25 eV relative binding energy shift between the UP and DW samples. This shift, along with the band dispersions, is consistent with the modest band bending near the surface and the reported ferroelectric-induced band offset across the first ≈1–2 nm near the surface. In the UP sample $\Delta\phi^{UP} = \phi_{PZT}\text{-}\phi_{Nb:STO} = 0.4$ eV, corresponding to upward band bending, and in the DW sample $\Delta\phi^{DW} = \phi_{PZT}\text{-}\phi_{LSMO} = -0.3$ eV associated with the downward band bending of PZT at the LSMO substrate [35]. A negatively charged LSMO interface stabilizes the ferroelectric state pointing toward the substrate (P−), while V(z) in **Figure 7b** depicts the positive charging into Nb:STO at the interface to correct the ferroelectric polarization pointing away from the substrate (P+) [35, 37–39, 42, 43]. A necessary prerequisite for the stabilization of the FE state appears to be the substrate-induced band bending at the bottom interface, which makes it easier for positive and negative charge sheets to accumulate at the surface/bottom interface and stablize the P+/P − FE polarization. **Figure 5d** shows the emergence of a new component in the Pb 4f spectra at lower binding energies, which suggests that the UP sample's creation of oxygen vacancies and the clustering of metallic Pb were caused by broken Pb-O bonds [41].

The fact that the DW sample lacks this component suggests that the oxygen vacancies and the corresponding negative charges that develop into the UP sample are necessary to offset the depolarizing field and stabilize the P+ FE state. They build up in the vicinity of the PZT surface to form a negative charge sheet that filters the polarization charges, which are fixed. However, in the DW sample with the P− FE state, the formation of cation vacancies (Pb and Ti) screens the depolarizing field produced by the fixed negative polarization charges at the PZT surface and positive at the bottom contact with LSMO.

4. Lead-based ferroelectric films: doped thin films

In this section, we describe our investigations related to the preparation of doped thin films, with Fe acting as acceptor and Nb acting as donor, and their contribution on understanding conduction type in ferroelectric films. In addition to doping studies, the fabrication of p-n junctions in ferroelectric materials is a topic of interest. Our studies revealed significant differences in electrical properties between Fe- and Nb-doped PZT films, including variations in coercive field, potential barrier height at the electrode interface, charge carrier concentration, and leakage current. These differences are attributed to the alterations introduced by the two dopants in the electronic properties of PZT. Furthermore, a change in doping type was found to affect polarization orientation, with upward polarization dominant in PZT-Nb and slightly downward polarization dominant in PZT-Fe. The establishment of p-n junctions in PZT films involves precise deposition of dopants known to behave as donors or acceptors when substituting Zr or Ti in the PZT lattice. These junctions offer novel device structures and functionalities, revealing quasi-linear current-voltage characteristics and temperature-dependent resistance and elucidating carrier injection mechanisms at electrode interfaces.

The PZT films were deposited using pulsed laser deposition (PLD) from custom-made targets containing precursor metal oxides with a purity of at least 99.99%.

Specifically, one target was doped with approximately 1% Fe (referred to as PZT-Fe), while the other was doped with approximately 1% Nb (PZT-Nb). The intended doping concentration was about 10^{20} cm^{-3} in the targets, although the actual doping level in the films might be lower. The deposition took place on single-crystal STO substrates with (001) orientation. First, a thin layer of $SrRuO_3$ (SRO), approximately 20 nm thick, was deposited by PLD to serve as the bottom electrode. The growth conditions for PZT films were similar to those detailed in a previous [44, 45]. Both single-doped PZT layers and bilayer p-n junctions were prepared in this study. Two types of bilayers were grown, resulting in the following structures: STO/SRO/PZT-Fe/PZT-Nb and STO/SRO/PZT-Nb/PZT-Fe. The thicknesses of the layers were equal, estimated to be approximately 200 nm and in the bilayer architecture to be 100 nm each. The thickness estimation was based on the number of laser pulses used during deposition.

4.1 Structural characterization

Figure 8 illustrates the outcomes of XRD structural analyses for simple PZT-doped layers, while **Figure 9** displays the XRD data for the bilayer structures. In all samples,

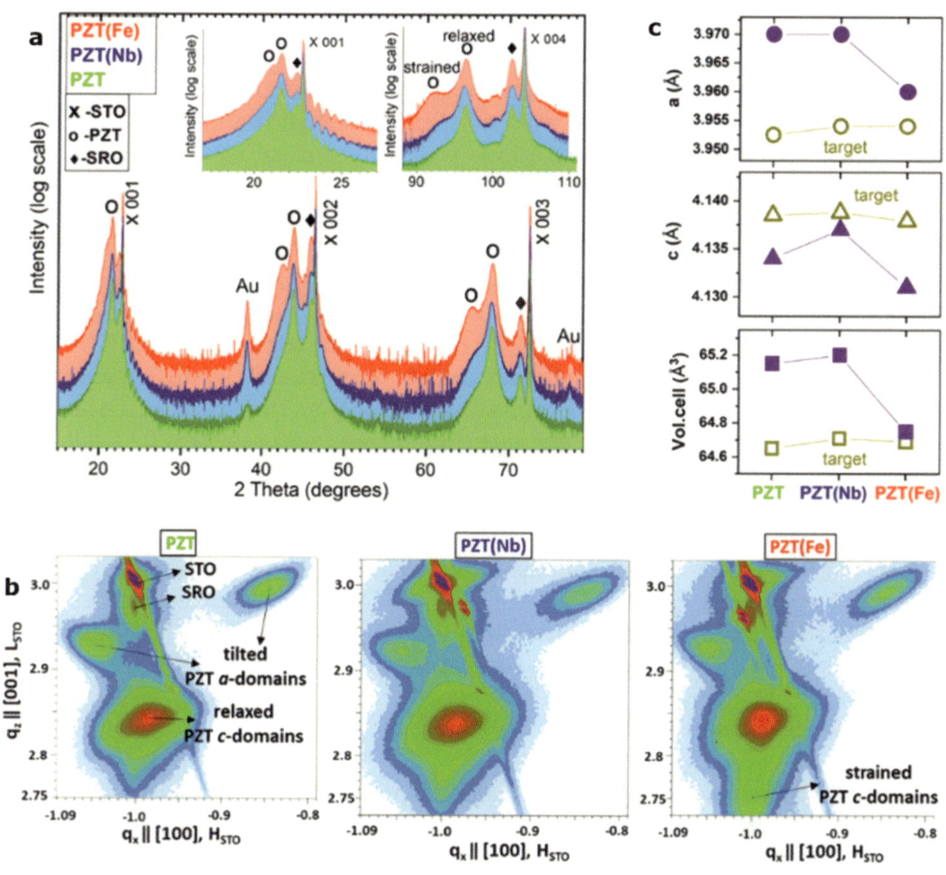

Figure 8.
(a) Diffractograms (2θ-ω scans) of single PZT layer samples, with details around lines 001 and 004 in inserts; (b) RSMs around nodes —103 of pseudocubic structures [45].

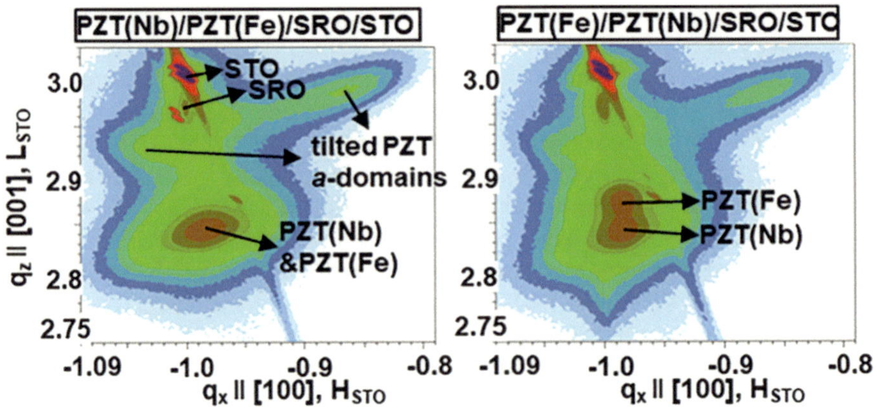

Figure 9.
RSMs around nodes—103 of PZT bilayer samples. (Reprint from Ref. [44] copyright 2023 American Chemical Society").

the structure primarily comprises PZT*c*-domains, where the *c* crystallographic axis of the tetragonal lattice is perpendicular to the film surface. Additionally, inclined PZT-domains are observed in the reciprocal space maps (**Figure 8b**), indicating relaxation through *a*-domains by forming the *a*/*c* structure.

This phenomenon is specific to PZT films with thicknesses exceeding 100 nm. Two distinct types of tetragonal PZT are discerned, characterized by different values of the lattice constant *c* perpendicular to the substrate. Most films exhibit a tetragonal structure with a *c* value close to the bulk (referred to as "relaxed"), while layers situated just above the substrate show a larger *c* value and in-plane lattice constant *a*, similar to that of the substrate (termed as "strained"). The proportion of the strained component varies depending on the sample. The analyses based on **Figure 8c**, reveal intriguing characteristics of the films; in contrast to expectations stemming from the lattice mismatch with the STO substrate, the out-of-plane parameter of the films decreases, while the in-plane parameter increases compared to the target.

Moreover, significant discrepancies are noted among the films, surpassing variations observed among the targets. Specifically, the PZT-Nb film exhibits the largest *c* lattice constant, whereas PZT-Fe displays the smallest; the un-doped PZT falls between these extremes, mirroring the trends observed in the films. Remarkably, the PZT-Fe film manifests a considerably smaller in-plane lattice constant compared to the others despite sharing the same *a* lattice constant as PZT-Nb. This observation suggests that Nb doping slightly increases the volume of the unit cell, while Fe doping reduces it relative to un-doped PZT. Such behavior can be rationalized by the inhibitory effect of Nb doping on the formation of oxygen vacancies, whereas Fe doping promotes their generation in the films compared to un-doped PZT.

4.2 Polarization switching at nanoscale

A specific poling map was selected to examine polarization switching at the nanoscale for simple PZT-doped layers, as depicted in **Figure 10a**. The voltage applied to the PFM tip was incrementally increased from −10 to 10 V along each scan line, as shown in **Figure 10b**. The written ferroelectric domains are visible in the piezoresponse phase images from **Figure 11**. In the as-grown un-doped PZT film, the polarization is predominantly upward-oriented (see **Figure 11f**) similar to previous findings [31].

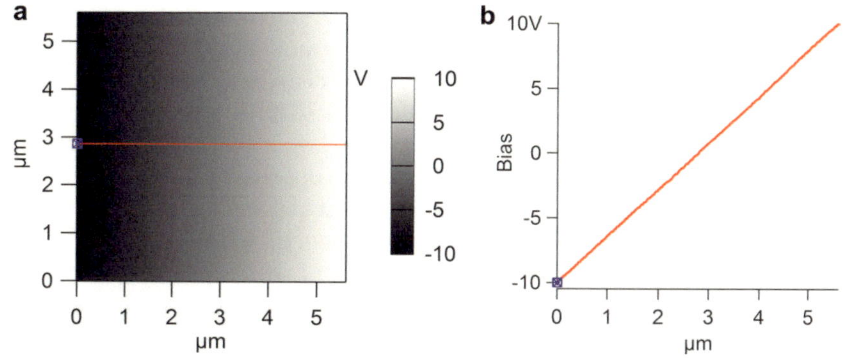

Figure 10.
(a) Poling map applied on a 6 × 6 μm surface for polarization switching; (b) section graph in the poling map image shows the applied voltage between −10 and 10 V [45].

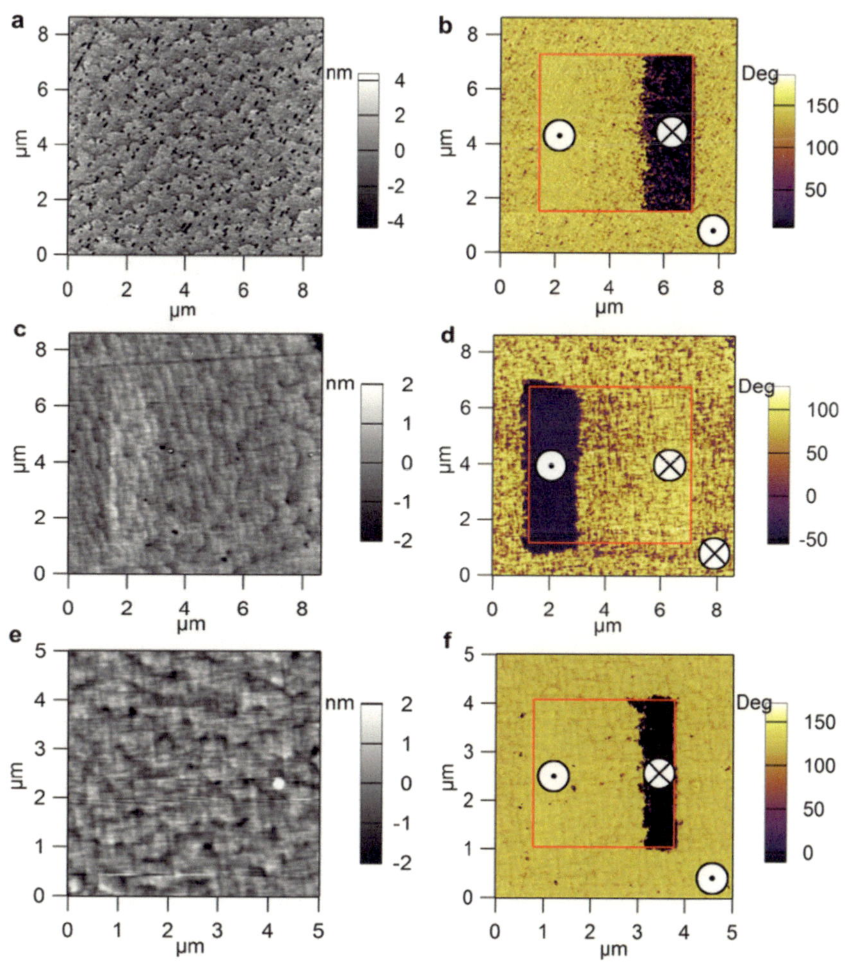

Figure 11.
AFM-PFM measurements on PZT-Nb (a), (b); PZT-Fe (c), (d); and un-doped PZT thin films (e), (f). Surface topography images (a), (c), (e) and piezoresponse phase images (b), (d), (f) after domain writing according to the polling map. The symbols used represent polarization orientation, with a solid circle inside a white circle indicating upward polarization and cross lines in a white circle representing downward polarization [45].

In the case of as-grown PZT-Nb, the dominant polarization orientation is upward, although some areas exhibit a downward polarization tendency. This deviation from un-doped PZT can be attributed to a reduction in the density of oxygen vacancies, influencing the compensation of the depolarization field during the growth process. This finding correlates with the lower leakage current and larger coercive field observed in PZT-Nb. For PZT-Fe, a mixture of domains with both upward and downward polarization is evident, with a larger area demonstrating downward polarization (estimated to about 80% of the as-grown area before poling, as shown in [45]).

This phenomenon can be attributed to Fe acting as an acceptor, facilitating the formation of oxygen vacancies for local charge compensation. Consequently, the distribution of Fe in the film and other structural defects influence the availability of charges for compensating the depolarization field, thereby favoring either upward or downward polarization orientation. Nonetheless, the transition from Nb (donor) to Fe (acceptor) doping visibly influences the PFM phase signal.

4.3 Electrical properties

The results of standard electrical measurements at room temperature are depicted in **Figure 12**. These measurements were conducted after establishing the same polarization state in the samples by applying a positive voltage on the top electrode

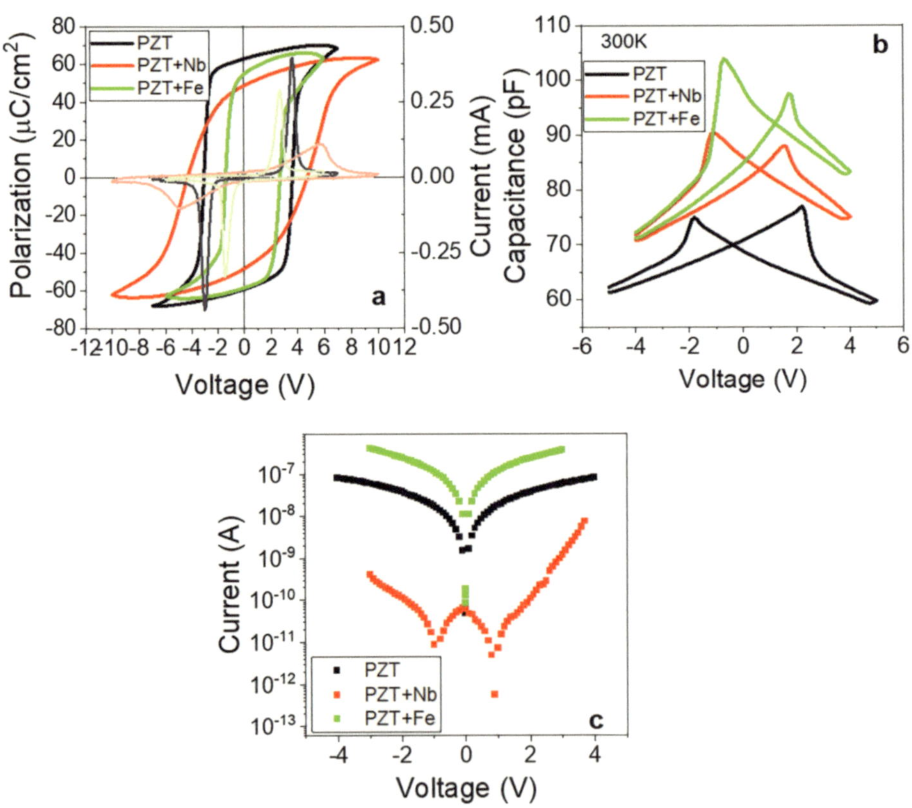

Figure 12.
The electrical characterization of the simple layers of PZT-Fe, PZT-Nb compared with the un-doped PZT. (a) Polarization-voltage loop, (b) capacitance-voltage characteristics, and (c) current-voltage characteristics presented in absolute value and in logarithmic scale [45].

for 10 seconds. Un-doped PZT exhibits intermediate polarization between PZT-Fe (lowest) and PZT-Nb (highest), likely due to differences in leakage current and lattice constants. Coercive field for un-doped PZT lies between PZT-Fe (lower) and PZT-Nb (higher), indicating varying ease of polarization switching attributed to leakage current. Internal electric field, though larger in doped samples, is insufficient alone to dictate polarization direction with its orientation consistent across films due to similar substrate conditions. Doped PZT films exhibit slightly higher dielectric constants, attributed to increased structural defects capable of responding to applied voltage signals. In paper [45], it can be also seen that the height of potential barriers at electrodes differs significantly, with un-doped PZT between PZT-Fe (lowest) and PZT-Nb (highest). Also, the effective density of charge carriers shows minimal variation among the three layers, with Fe doping promoting oxygen vacancies formation and Nb doping inhibiting it. Fe introduces an acceptor level near the valence band, while Nb introduces a donor level near the conduction band (details in [45]).

PZT-Nb displays n-type behavior, while PZT-Fe shows slight p-type behavior. Similar effective densities of free carriers correspond to remnant polarization values, indicating polarization's control over these densities, regardless of doping.

Self-doping, likely *via* oxygen vacancies, regulates carrier densities, with consistent contributions from SRO electrodes. Differences in barrier height stem from Fermi level positioning: un-doped PZT has n-type behavior due to oxygen vacancies, placing the Fermi level near the conduction band. PZT-Nb's Fermi level is influenced by donor levels, leading to higher potential barriers. In PZT-Fe, the Fermi level is near the valence band, resulting in low barriers. Variations in barrier height affect leakage current magnitude, with PZT-Fe exhibiting significantly higher leakage than PZT-Nb. Modest Fe and Nb doping induces notable differences in macroscopic properties of epitaxial PZT films. These differences arise from distinct electronic properties induced by Fe and Nb doping. The effects of p-type and n-type doping can be discerned through specialized poling procedures, offering potential for ferroelectric p-n homojunctions.

Hysteresis measurements reveal distinct hysteresis loops for both bilayers, PZT-Nb/PZT-Fe and PZT-Fe/PZT-Nb, with notable differences in loop shape, magnitude, remnant polarizations, and coercive voltages (**Figure 13**). When PZT-Fe is the first deposited layer, higher polarization and coercive voltage values are observed compared to when PZT-Nb is the first layer. This variance can be attributed to differing electrostatic conditions at the interface with the bottom SRO electrode. As presented above and in [45], the estimated potential barrier is larger for the PZT-Nb/SRO interface than for the PZT-Fe/SRO interface, approximately 0.3 eV compared to 0.1 eV. This distinction can affect the compensation of the depolarization field during both the growth process and polarization switching. A larger potential barrier impedes the flow of charges involved in compensating the depolarization field, resulting in lower polarization values. The I-V characteristics obtained and presented in **Figure 13b** display an unexpectedly symmetric linear shape despite the differing potential barriers at the two interfaces with the bottom and top SRO electrodes. This contradicts the anticipated presence of Schottky-like contacts at the electrode interfaces in bilayer structures. The unusual shape of the I-V characteristics may be attributed to two factors: (i) the coexistence of two types of carriers (electrons and holes) in the p-n homojunction and (ii) complex charge compensation processes occurring near the contacts and at the PZT-Nb/PZT-Fe interface. Additionally, it is noted that the PZT-Fe/PZT-Nb/SRO/STO structure exhibits lower resistance than the PZT-Nb/PZT-Fe/SRO/STO structure [44].

The slope of the linear I-V characteristic represents the electrical conductance or the inverse of resistance. Resistance values were extracted from I-V measurements

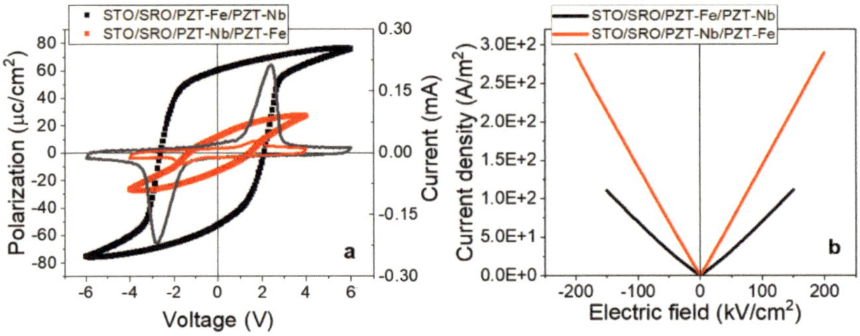

Figure 13.
(a) The hysteresis measurements for the bilayered structures. (b) The current-voltage characteristics for the bilayered structures (Reprint from Ref. [44] copyright 2023 American Chemical Society).

at different temperatures, and the observed decrease in resistance at 300 K is approximately four orders of magnitude, indicating semiconductor behavior as presented in [44]. Arrhenius plots were employed to extract an activation energy for conduction in the two bilayer structures. The obtained values fall within the range of 0.13–0.17 eV, which lies between the estimated potential barrier heights at the SRO electrodes (around 0.1 eV at the PZT-Fe/SRO interface and 0.3 eV at the PZT-Nb/SRO interface, as estimated for single-phase capacitor structures).

5. Lead-free thin films

In this section, we present our investigations related to the preparation of lead-free ferroelectric thin films and their properties on energy storage domain. The hysteresis polarization loops analysis is employed in this study to deliver useful information on the energy storage characteristics of lead-free ferroelectric capacitors based on $0.92(Na_{0.5}Bi_{0.5})TiO_3$-$0.08BaTiO_3$ (BNT-BT). The total stored charge and efficiency measured on a different range of temperatures, from 100 to 400 K, indicating an excellent thermal stability. Using an in-house obtained BNT-BT target, an epitaxial thin film was prepared for this investigation on a STO (001) substrate with SRO bottom electrode. More information regarding target preparation is provided in [46]; for the SRO bottom electrode deposition, we always use the same recipe, see above. The oxygen atmosphere, the substrate temperature, and post-annealing conditions were chosen to promote the growth of high-quality BNT-BT film with good crystallinity and stoichiometry. The thin film deposition occurs in an oxygen environment at a pressure of 0.2 mbar. The target is ablated using a laser fluence of 1 J/cm^2 with a pulse frequency of 10 Hz. The substrate where the thin film is placed is heated to around 600°C. This temperature is used to promote optimum crystallization of the deposited material in addition to preventing the formation of Bi and Na vacancies caused by their volatility. After deposition, the film was gradually cooled on high oxygen pressure to room temperature.

5.1 Structural characterization

XRD measurements were performed using a Rigaku-SmartLab X-ray diffractometer (Rigaku Corporation, Tokyo, Japan) with a conventional Cu anode X-ray tube, powered at 40 kV and 40 mA. The symmetric XRD scans corrected for the single-crystal surface miscut, 2θ ω scans, were performed with a Ge (220) monochromator in the incident beam and a HyPix detector in 0D mode, in the 2θ range 10–57°. The reciprocal space mapping (RSM) images were taken with a HyPix 3000 camera in 2D mode, around χ = 0 (symmetric to the sample surface), within 2θ = 35–51°, and χ = ±15°. The BNT-BT film is stabilized in tetragonal structure with highly *c* axis texture, as deduced from the 2Θ-ω scan shown in **Figure 14a**. The SRO layer's layer fringes are visible in the inset of **Figure 14a**, indicating the atomic scale level smoothness of its sides. The film thickness calculated from the layer fringes' period is about 20 nm. Reciprocal space maps (**Figure 14c**) show two distinct types of *c*-domains with varying lattice constant *c*-perpendicular values on the substrate. The layers immediately above the substrate have a larger *c* value and an in-plane lattice constant *a*, close to the substrate one (named here "strained"), while the majority of the films have a tetragonal structure with a *c* value close to the bulk one (named here "relaxed") [47]. The strained layer should be related to the thin layer visible in HRTEM images at the SRO-BNT-BT interface (**Figure 15a**). The in-plane orientation of the films has been analyzed performing azimuth scans in asymmetric geometry measurement locating tilt planes to the film surface.

The {103} planes were used for STO and SRO, and the {113} plane family for the assumed tetragonal BNT film and were scanned over a whole rotation around the surface normal (**Figure 15b**). The relative positions of the phi-scan maxima of the three structures indicate that the square edge of BNT is parallel to the face diagonal of SRO or STO: BNT [100] // SRO [110] //STO [110]. This in-plain orientation is determined by the almost perfect matching of the periods in these directions: *a*BNT ≈ *a*STO.

Cross-section transmission electron microscopy (TEM) specimen of BNT-BT has been prepared from the as-deposited sample by mechanical polishing down to ca. 30 μm, followed by ion milling in a Gatan PIPS machine at 4 kV accelerating voltage and 7° incidence angle. Low-voltage (2 kV) milling was used as the final ion polishing stage in order to reduce the amorphous surface layer enveloping the specimen. TEM observations have been performed using a probe-corrected analytical high-resolution JEM ARM 200F electron microscope operated at 200 kV. The conventional TEM image in **Figure 15** a gives a general cross-section view of the BNT-BT/SRO/STO film showing the growth morphology and the layer's distribution on the (100) STO substrate. The SRO layer is visible as a narrow band of uniform contrast and thickness (20 nm) on the top of the STO substrate. The BNT-BT layer has a thickness of 250 nm with a surface roughness below 5 nm, as displayed by the cross-section image.

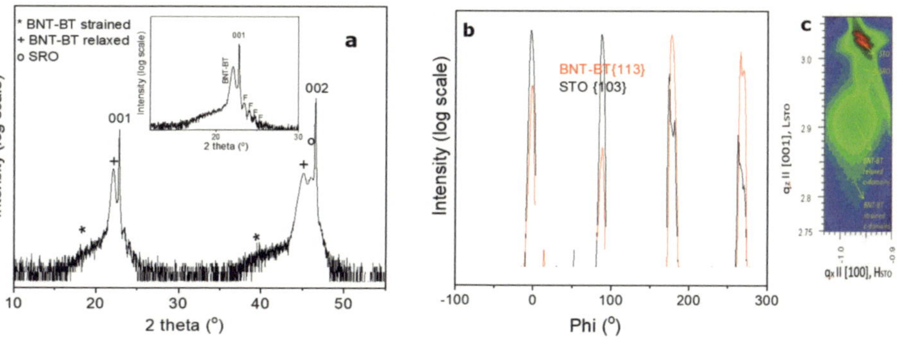

Figure 14.
(a) XRD 2Θ-ω scan; (b) phi-scans recorded in asymmetric geometry for the STO substrate, the SRO, and film and BNT films; (c) RSMs around nodes {103} of pseudocubic structures.

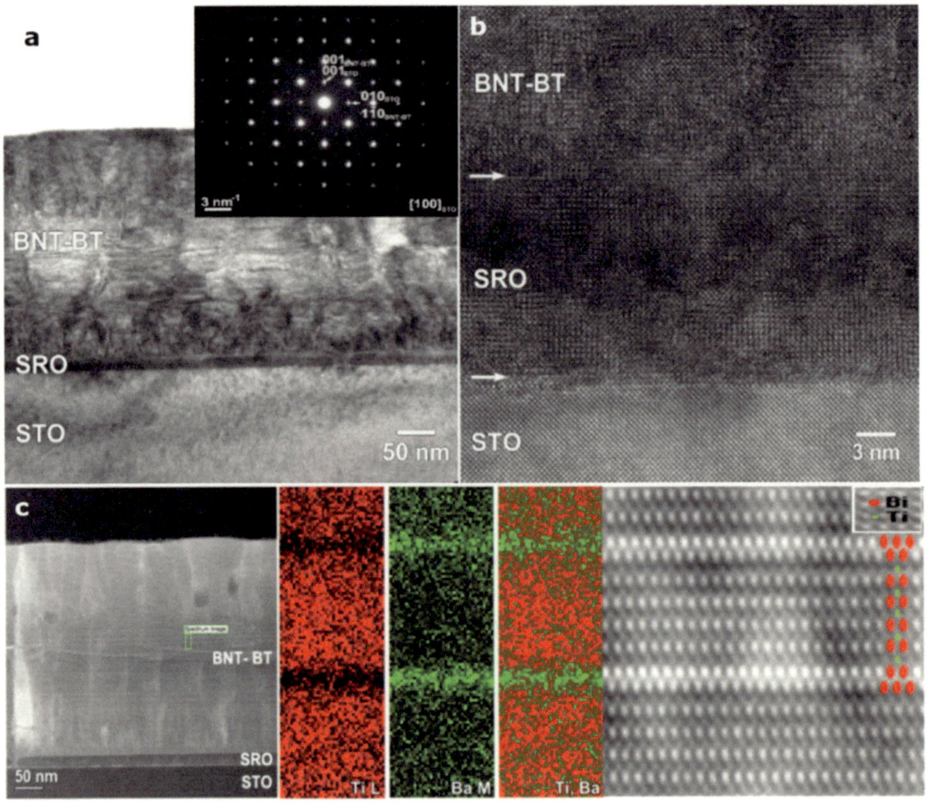

Figure 15.
(a) Cross-section TEM image at low-magnification showing BNT-BT/SRO/STO film, inset: SAED pattern corresponding to the TEM image; (b) HRTEM image at the BNT-BT/SRO and SRO/STO interfaces; (c) periodical defects observed in STEM image analyzed by EELS-SI.

A strong diffraction contrast can be observed across the BNT-BT layer either parallel or perpendicular to the growth direction, attesting the elastic strain due to various growth defects. The epitaxial growth of the BNT-BT and SRO layers onto the STO substrate is demonstrated by the selected-area electron diffraction (SAED) pattern in the inset of **Figure 15a**, where the STO substrate is in [100] zone axis orientation. The diffraction pattern (DP) contains contributions from the substrate, the SRO electrode, and the BNT-BT layer. A single family of spots is visible, the 1.3% lattice mismatch being too small to produce evident spot splitting, especially for the low-index spots. The observable deformation of the diffraction spots, which can be noticed as streaks parallel to the (001) STO and (010) STO reciprocal vectors, is most probably related to the high density of defects in BNT-BT oriented perpendicular to the interface (next to the interface) or parallel to it (away from the interface). The diffraction spots attributed to BNT-BT have been indexed according to the tetragonal structure of BNT-BT in agreement with the phase diagram for this stoichiometry at room temperature [47]. **Figure 15c** presents the cross-sectional HAADF-STEM image at low magnification of the BNT-BT/SRO/STO heterostructure where the spectrum image for EFLS was selected, followed by a relative composition maps in false colors for Ti L and Ba M extracted from EELS-SI data cube and RGB composed image by overlapping the Ti and Ba elemental maps. The Aurivillius phase was also identified, and their structure is built by alternating layers of [Bi_2O_2]2+ and pseudo-perovskite blocks, with perovskite layers that are n (5 in our case) octahedral layers in thickness [48, 49]. In TEM images recorded on the BNT-BT thin film, the presence of different types of defects such as Aurivillius phase, Barium precipitates, grain boundaries, and dislocations was observed.

5.2 Stored/released energies and the efficiency of BNT-BT

The physical properties of a BNT-BT capacitor, such as the area of the electrodes, thickness, dielectric constant, and leakage, determine the amount of charge that it can store and, thus, its capacity. In **Figure 16**, a PUND hysteresis measure at 300 K is presented. The maximum polarization at the maximum voltage applied is 45 μC/cm^2.

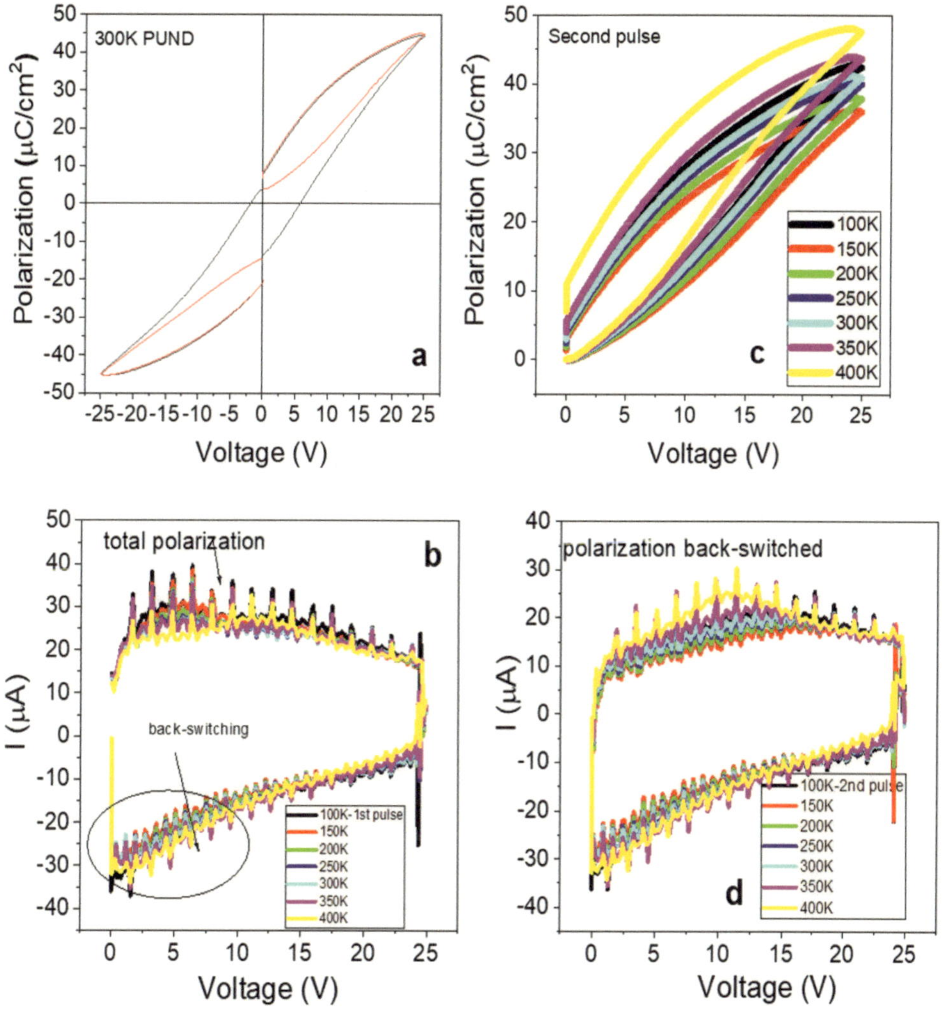

Figure 16.
(a) Polarization-voltage characteristics obtained with PUND measurement. The back loops represent the results from the first and the third pulse, while the red loops represent the results from the second and the fourth pulse, (b) the current-voltage characteristics obtained during the first pulse for different voltages, (c) the polarization-voltage loop measured obtained during the second positive pulse for different temperature, and (d) the current-voltage characteristics obtained after the second positive pulse for different temperatures.

But, the remnant polarization is 10 μC/cm². This represents the capacitor based on BNT-BT thin film that presents a significant back-switching of polarization when the voltage drops to zero. This property can be used for the application of storage and fast discharging of electrostatic energy in supercapacitors. **Figure 16c** represents the modification of polarization-voltage characteristics obtained for second consecutive positive voltage pulse for temperature from 100 to 400 K. As can be seen in **Figure 16b** and **d**, the total polarization, represented here as the area under the switching peak from **Figure 16b**, has a small decreasing with temperature, but the back-switching component is increasing with temperature.

Thus, the polarization measured by the second consecutive positive pulse increases with temperature as is explicitly represented in **Figure 16c** in polarization-voltage loops or in **Figure 16d** by current-voltage loops.

From characteristics presented in **Figure 16**, the stored and released density of energy and also the efficiency are deduced using the formulas:

$$E_{charge} = \int_{0}^{Pmax} EdP \tag{1}$$

$$E_{discharge} = \int_{Pmax}^{Prem} EdP \tag{2}$$

$$\eta = \frac{E_{discharge}}{E_{charge}} \tag{3}$$

The stored and released energies and also the efficiency for temperatures between 100 and 400 K are represented in **Figure 17a**, for the first pulse, and also in **Figure 17b**, for only the second pulse, which represents the back-switching. The total stored charge is between 25 and 22 J/cm³ for the first pulse and around 22 and 17 J/cm³ for the second pulse. The discharged energy presents a small increasing with temperature between 10 and 11 J/cm³ for both first and second pulse. Thus, efficiency also presents a small increasing by increasing the temperature. Around 60% efficiency is obtained for the second pulse and smaller value of 45% for the first pulse.

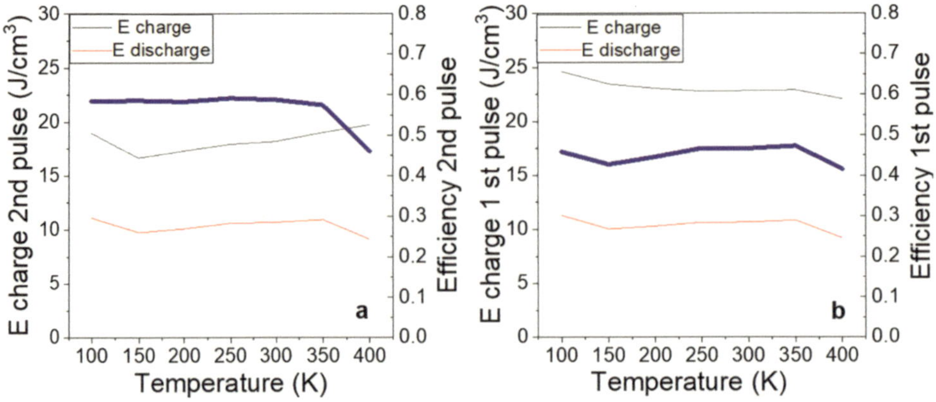

Figure 17.
The stability of the stored and released energies and also the effiencies for temperatures between 100 and 400 K; (a) specific for the first pulse; (b) specific for the second pulse.

In comparison to earlier publications, the E charge and E discharge values achieved in this work are higher than those found on bulk ceramics or polycrystalline films [6–9]. We also found that the characteristics of epitaxial BNT-BT thin films remain constant across a wide temperature range from 100 to 350 K.

6. Conclusions

This chapter presents the results found in epitaxial ferroelectric materials, either lead-based or lead-free, obtained by the pulsed laser deposition technique. An overview of the deposition and growth strategies employed to achieve high-quality epitaxial ferroelectric structures with well-defined interfaces is provided initially. Developing lead-based epitaxial thin films with variable thickness and their role in clarifying the fundamental issues of ferroelectric thin films are presented. In the final section of this chapter, we describe the growth of a lead-free epitaxial thin film, followed by its characteristics in the energy storage domain.

Acknowledgements

This research was funded by Core Program, component project PN23080202 and PN23080303, funded by Romanian Ministry of Research, Innovation and Digitization. The fee for open access publication was supported from the project 35PFE/2021, funded by the Romanian Ministry of Research, Innovation and Digitization. D. Popescu acknowledges the received funding from the PN-III-P1-1.1-TE-2021-0136/TE-2021-1053/RaBit.

Conflict of interest

The authors declare no conflict of interest.

Author details

Cristina Chirila*, Andra G. Boni, Lucian D. Filip, Mihaela Botea, Dana Popescu, Viorica Stancu, Lucian Trupina, Luminita Hrib, Raluca Negrea Ioana Pintilie and Lucian Pintilie
National Institute of Materials Physics, Bucharest-Magurele, Romania

*Address all correspondence to: dragoi@infim.ro

References

[1] Bauer S, Rodrigues A, Baumbach T. Real time in situ x-ray diffraction study of the crystalline structure modification of $Ba_{0.5}Sr_{0.5}TiO_3$ during the post-annealing. Scientific Reports. 2018;**8**:11969. DOI: 10.1038/s41598-018-30392-y

[2] Martin LW, Chu Y-H, Ramesh R. Advances in the growth and characterization of magnetic, ferroelectric, and multiferroic oxide thin films. Materials Science & Engineering R: Reports. 2010;**68**:89-133. DOI: 10.1016/j.mser.2010.03.001

[3] Ramesh V. Interface effects in epitaxial multiferroic complex oxide heterostructures [thesis]. UNSW Sydney. 2019. DOI: 10.26190/unsworks/21132

[4] Koster G, Blank DHA, Rijnders GAJHM. Oxygen in complex oxide thin films grown by pulsed laser deposition: A perspective. Journal of Superconductivity and Novel Magnetism. 2020;**33**:205-212. DOI: 10.1007/s10948-019-05276-5

[5] Tyunina M, Pacherova O, Peräntie J, Savinov M, Jelinek M, Jantunen H, et al. Perovskite ferroelectric tuned by thermal strain. Scientific Reports. 2019;**9**:3677. DOI: 10.1038/s41598-019-40260-y

[6] Cao WP, Li WL, Dai XF, Zhang TD, Sheng J, Hou YF, et al. Large electrocaloric response and high energy-storage properties over a broad temperature range in lead-free NBT-ST ceramics. Journal of the European Ceramic Society. 2016;**36**:593-600. DOI: 10.1016/j.jeurceramsoc.2015.10.019

[7] Zhao Y, Hao X, Li M. Dielectric properties and energy-storage performance of $(Na_{0.5}Bi_{0.5})TiO_3$ thick films. Journal of Alloys and Compounds. 2014;**601**:112-115. DOI: 10.1016/j.jallcom.2014.02.137

[8] Li J, Yang H, Lin Y, Gan X. High energy storage property of $Na_{0.5}Bi_{0.5}TiO_3$-based ceramics modified by complex-ion and $SrTiO_3$. Journal of Alloys and Compounds. 2022;**909**:164579. DOI: 10.1016/j.jallcom.2022.164579

[9] Bousquet M, Batista L, Dellis JL, Boulle A, Rabe U, Durand-Drouhin O, et al. Structural and electrical properties of $Bi_{0.5}Na_{0.5}TiO_3$ based superlattices grown by pulsed laser deposition. Journal of Applied Physics. 2014;**116**:194104. DOI: 10.1063/1.4901931

[10] Dewan S, Khanikar PD, Mudgal R, Singh A, Muduli PK, Singh R, et al. Large-area GeSe realized using pulsed laser deposition for ultralow-noise and ultrafast broadband phototransistors. ACS Applied Materials & Interfaces. 2023;**15**:27285-27298. DOI: 10.1021/acsami.3c02522

[11] Sakai S, Takahashi M, Motohashi K, Yamaguchi Y, Yui N, Kobayashi T. Large-area pulsed-laser deposition of dielectric and ferroelectric thin films. Journal of Vacuum Science and Technology A. 2007;**25**:903-907. DOI: 10.1116/1.2748808

[12] Jamaludin N, Chaudhary KT, Haider Z, Duralim M, Ismail FD, Roslan MS, et al. Effect of laser energy and wavelength on average size of gold nanoparticles synthesized by pulsed laser ablation in deionized water. Journal of Physics Conference Series. 2020;**1484**:012029. DOI: 10.1088/1742-6596/1484/1/012029

[13] Singh RK, Narayan J. A novel method for simulating laser-solid

interactions in semiconductors and layered structures. Materials Science and Engineering B. 1989;**3**:217-230. DOI: 10.1016/0921-5107(89)90014-7

[14] Haider AJ, Alawsi T, Haider MJ, Taha BA, Marhoon HA. A comprehensive review on pulsed laser deposition technique to effective nanostructure production: Trends and challenges. Optical and Quantum Electronics. 2022;**54**:488. DOI: 10.1007/s11082-022-03786-6

[15] Shepelin NA, Tehrani ZP, Ohannessian N, Schneider CW, Pergolesi D, Lippert T. A practical guide to pulsed laser deposition. Chemical Society Reviews. 2023;**52**:2294-2321. DOI: 10.1039/D2CS00938B

[16] García López J, Blank DHA, Rogalla H, Siejka J. Role of the oxygen plasma during in situ growth of $YBa_2Cu_3O_{6+x}$ thin films by pulsed laser deposition. Physica C: Superconductivity. 1998;**307**:298-306. DOI: 10.1016/S0921-4534(98)00534-6

[17] Takahashi T, Tani S, Kuroda R, Kobayashi Y. Precision measurement of ablation thresholds with variable pulse duration laser. Applied Physics A: Materials Science & Processing. 2020;**126**:582. DOI: 10.1007/s00339-020-03754-5

[18] Galca AC, Stancu V, Husanu MA, Dragoi C, Gheorghe NG, Trupina L, et al. Substrate–target distance dependence of structural and optical properties in case of Pb(Zr,Ti)O3 films obtained by pulsed laser deposition. Applied Surface Science. 2011;**257**:5938-5943. DOI: 10.1016/j.apsusc.2011.01.056

[19] Gabriel V, Kocán P, Bauer S, Nergis B, Rodrigues A, Horák L, et al. Effect of pulse laser frequency on PLD growth of LuFeO3 explained by kinetic simulations of in-situ diffracted intensities. Scientific Reports. 2022;**12**:5647. DOI: 10.1038/s41598-022-09414-3

[20] Schou J. Physical aspects of the pulsed laser deposition technique: The stoichiometric transfer of material from target to film. Applied Surface Science. 2009;**255**:5191-5198. DOI: 10.1016/j.apsusc.2008.10.101

[21] Haile HT, Dejene FB. Influence of oxygen deposition pressure on the structural, morphological, photoluminescence and thermoluminescence properties of Y2SiO5:Ce3+ by pulsed laser deposition method. Inorganic Chemistry Communications. 2021;**127**:108554. DOI: 10.1016/j.inoche.2021.108554

[22] Zhou Y, Wang D, Li Y, Jing L, Li S, Chen X, et al. Critical effect of oxygen pressure in pulsed laser deposition for room temperature and high performance amorphous In-Ga-Zn-O thin film transistors. Nanomaterials. 2022;**12**:4358. DOI: 10.3390/nano12244358

[23] Uecker R, Bertram R, Brützam M, Galazka Z, Gesing TM, Guguschev C, et al. Large-lattice-parameter perovskite single-crystal substrates. Journal of Crystal Growth. 2017;**457**:137-142. DOI: 10.1016/j.jcrysgro.2016.03.014

[24] Schlom DG, Chen LQ, Fennie CJ, et al. Elastic strain engineering of ferroic oxides. MRS Bulletin; 2014;**39**:118-130. DOI: 10.1557/mrs.2014.1

[25] Bachelet R, Sánchez F, Palomares FJ, Ocal C, Fontcuberta J. Atomically flat SrO-terminated $SrTiO_3$(001) substrate. Applied Physics Letters. 2009;**95**:141915. DOI: 10.1063/1.3240869

[26] Phark S, Chang YJ. Nucleation and growth of primary nanostructures in $SrTiO_3$ homoepitaxy. Nanoscale Research

Letters. 2015;**10**:80. DOI: 10.1186/s11671-015-0805-7

[27] Huijbregtse JM, Rector JH, Dam B. Effect of the two (1 0 0) $SrTiO_3$ substrate terminations on the nucleation and growth of $YBa_2Cu_3O_{7-\delta}$ thin films. Physica C: Superconductivity. 2001;**351**:183-199. DOI: 10.1016/S0921-4534(00)01616-6

[28] Crespo A, Gallenberger J, De Santis M, Langlais V, Carla F, Caicedo JM, et al. Heteroepitaxial growth of anatase (0 0 1) films on SrTiO3 (0 0 1) by PLD and MBE. Applied Surface Science. 2023;**632**:157586. DOI: 10.1016/j.apsusc.2023.157586

[29] Thune E, Fakih A, Matringe C, Babonneau D, Guinebretière R. Understanding of one dimensional ordering mechanisms at the (001) sapphire vicinal surface. Journal of Applied Physics. 2017;**121**:015301. DOI: 10.1063/1.4973341

[30] Luo Y, Li X, Chang L, Gao W, Yuan G, Yin J, et al. Upward ferroelectric self-poling in (001) oriented $PbZr_{0.2}Ti_{0.8}O_3$ epitaxial films with compressive strain. AIP Advances. 2013;**3**:122101. DOI: 10.1063/1.4840595

[31] Pintilie L, Ghica C, Teodorescu CM, Pintilie I, Chirila C, Pasuk I, et al. Polarization induced self-doping in epitaxial $Pb(Zr_{0.20}Ti_{0.80})O_3$ thin films. Scientific Reports. 2015;**5**:14974. DOI: 10.1038/srep14974

[32] Yan K, Fan R, Chen M, Sun K, Yin L, Li H, et al. Perovskite $(La,Sr)MnO_3$ with tunable electrical properties by the Sr-doping effect. Journal of Alloys and Compounds. 2015;**628**:429-432. DOI: 10.1016/j.jallcom.2014.12.137

[33] Greculeasa SG, Stanciu A-E, Leca A, Kuncser A, Hrib L, Chirila C, et al. Influence of thickness on the magnetic and Magnetotransport properties of epitaxial $La_{0.7}Sr_{0.3}MnO_3$ films deposited on STO (0 0 1). Nanomaterials. 2021;**11**:3389. DOI: 10.3390/nano11123389

[34] Juan D, Pruneda M, Ferrari V. Localized electronic vacancy level and its effect on the properties of doped manganites. Scientific Reports. 2021;**11**:6706. DOI: 10.1038/s41598-021-85945-5

[35] Popescu DG, Husanu MA, Constantinou PC, Filip LD, Trupina L, Bucur CI, et al. Experimental band structure of $Pb(Zr,Ti)O_3$: Mechanism of ferroelectric stabilization. Advancement of Science. 2023;**10**:2205476. DOI: 10.1002/advs.202205476

[36] Pasuk I, Neațu F, Neațu Ș, Florea M, Istrate CM, Pintilie I, et al. Structural details of $BaTiO_3$ nano-powders deduced from the anisotropic XRD peak broadening. Nanomaterials. 2021;**11**:1121. DOI: 10.3390/nano11051121

[37] Tănase LC, Abramiuc LE, Popescu DG, Trandafir A-M, Apostol NG, Bucur IC, et al. Polarization orientation in lead zirconate titanate (001) thin films driven by the interface with the substrate. Physical Review Applied. 2018;**10**:034020. DOI: 10.1103/PhysRevApplied.10.034020

[38] Popescu DG, Husanu MA, Chirila C, Pintilie L, Teodorescu CM. The interplay of work function and polarization state at the Schottky barriers height for Cu/$BaTiO_3$ interface. Applied Surface Science. 2020;**502**:144101. DOI: 10.1016/j.apsusc.2019.144101

[39] Popescu DG, Husanu MA, Chirila C, Pintilie L, Teodorescu CM. Impact on ferroelectricity and band alignment of gradually grown Au on $BaTiO_3$. Physica Status Solidi (RRL) – Rapid Research

Letters. 2019;**13**:1900077. DOI: 10.1002/pssr.201900077

[40] Bocirnea AE, Popescu DG, Chirila C, Costescu RM, Kuncser V, Stancu V, et al. Polarization-dependent magnetism of the Ni/$BaTiO_3$ interface. Physical Review Materials. 2020;**4**:034402. DOI: 10.1103/PhysRevMaterials.4.034402

[41] Popescu DG, Huşanu MA, Trupină L, Hrib L, Pintilie L, Barinov A, et al. Spectro-microscopic photoemission evidence of charge uncompensated areas in Pb(Zr,Ti)O_3(001) layers. Physical Chemistry Chemical Physics. 2014;**17**:509-520. DOI: 10.1039/C4CP04546G

[42] Apostol NG, Stoflea LE, Lungu GA, Tache CA, Popescu DG, Pintilie L, et al. Band bending at free Pb(Zr,Ti)O_3 surfaces analyzed by X-ray photoelectron spectroscopy. Materials Science and Engineering B. 2013;**178**:1317-1322. DOI: 10.1016/j.mseb.2013.02.007

[43] Popescu DG, Barrett N, Chirila C, Pasuk I, Husanu MA. Influence of hole depletion and depolarizing field on the $BaTiO_3/La_{0.6}Sr_{0.4}MnO_3$ interface electronic structure revealed by photoelectron spectroscopy and first-principles calculations. Physical Review B. 2015;**92**:235442. DOI: 10.1103/PhysRevB.92.235442

[44] Boni A-G, Chirila C, Trupina L, Radu C, Filip LD, Moldoveanu V, et al. Resistive-like behavior of ferroelectric p–n bilayer structures based on epitaxial $Pb(Zr_{0.2}Ti_{0.8})O_3$ thin films. ACS Applied Electronic Materials. 2023;**5**:957-967. DOI: 10.1021/acsaelm.2c01497

[45] Chirila CF, Stancu V, Boni GA, Pasuk I, Trupina L, Filip LD, et al. Controlling polarization direction in epitaxial $Pb(Zr_{0.2}Ti_{0.8})O_3$ films through Nb (n-type) and Fe (p-type) doping. Scientific Reports. 2022;**12**:755. DOI: 10.1038/s41598-022-04802-1

[46] Dragoi C, Cernea M, Trupina L. Lead-free ferroelectric $BaTiO_3$ doped-$(Na_{0.5}Bi_{0.5})TiO_3$ thin films processed by pulsed laser deposition technique. Applied Surface Science. 2011;**257**:9600-9605. DOI: 10.1016/j.apsusc.2011.06.075

[47] Ogino M, Noguchi Y, Kitanaka Y, Miyayama M, Moriyoshi C, Kuroiwa Y. Polarization rotation and monoclinic distortion in ferroelectric $(Bi_{0.5}Na_{0.5})TiO_3$–$BaTiO_3$ single crystals under electric fields. Crystals. 2014;**4**:273-295. DOI: 10.3390/cryst4030273

[48] Choi WS, Seo SSA, Lee HN. 12 - optoelectronics: Optical properties and electronic structures of complex metal oxides. In: Koster G, Huijben M, Rijnders G, editors. Epitaxial Growth Complex Met. Oxides. Cambridge, United Kingdom: Woodhead Publishing; 2015. pp. 331-363. DOI: 10.1016/B978-1-78242-245-7.00012-9

[49] Zulhadjri Z, Sahiga JI, Wendari T, Emriadi E. Synthesis and structural analysis of aurivillius phase, Ca1-xBi3+xNdTi4-xMnxO15. Research Journal of Chemistry and Environment. 2018;**22**:4-9

Chapter 4

Synthesis of Nanoparticles Using Pulsed Laser

Hamed Naderi-Samani and Reza Shoja Razavi

Abstract

This chapter provides an extensive discussion of the pulsed laser ablation in liquid (PLAL) method for synthesizing nanoparticles. It covers the production of various types of nanoparticles, such as metal, semiconductor, and metal-oxide nanoparticles, and the impact of laser parameters on their properties, such as size, shape, composition, and crystallinity. The chapter also delves into the physical and chemical processes involved in PLAL, including nucleation, growth, and coalescence, and how they can be controlled to achieve tailored nanoparticle synthesis. Additionally, it examines the challenges and limitations of PLAL, such as particle aggregation, contamination, and reproducibility, and strategies for improving nanoparticle stability and dispersibility. This chapter is a valuable resource for researchers and scientists in the laser synthesis of nanoparticles, emphasizing the significance of pulsed laser parameters in achieving desired nanoparticle properties.

Keywords: nanoparticles, laser synthesis, pulsed laser ablation in liquids, laser pulse parameters, size control

1. Introduction

Pulsed laser ablation in liquid (PLAL) is divided into top-down methods in the synthesis of nanoparticles. The PLAL method is performed under ambient conditions and does not require temperature or pressure. Nanoparticles can be synthesized using the PLAL process, which has an almost unlimited domain. The laser source is known as the most significant parameter in the synthesis of nanoparticles by the PLAL. In most of the research conducted on the synthesis of nanoparticles by the PLAL method, nanosecond (ns) pulsed lasers have been applied. Millisecond (ms), microsecond (μs), femtosecond (fs), and picosecond (ps) pulsed lasers have also been used by researchers in the synthesis of nanoparticles. Lasers with different pulse widths due to variable interactions with liquids and targets give researchers various nanostructure synthesis choices.

Liquid is the second key parameter in the PLAL process after the laser source. In various research, liquids such as water, alcohol, acetone, sodium dodecyl sulfate (SDS), polymers, liquid nitrogen, liquid helium, have been used for the synthesis of nanostructures, which has expanded the application of the PLAL process in various fields. Furthermore, the research examined the effect of external factors on the PLAL process using electric field-assisted pulsed laser ablation in liquid (EF-PLAL) and

 IntechOpen

electrochemistry-assisted laser ablation in liquid (EC-PLAL). It has been shown that changing these external factors can easily alter nanomaterials' morphology, synthesis, and structure. Nanostructure morphology and synthesis can be controlled by considering the above three factors (laser, liquid, and external factors) in the PLAL process.

This part reviews the study's references and examines the PLAL process's mechanisms, the laser's effect, the ablation environment, and target composition parameters. Finally, the conclusions in this field are provided.

2. Synthesis of nanoparticles by laser

Over the past few decades, a new field of research called laser synthesis of colloids (LSC) has been created to produce nanoparticles [1]. According to **Figure 1**, LSC can be classified by four different methods: laser ablation in liquid (LAL), laser fragmentation in liquid (LFL), laser melting in liquid (LML), and laser defect-engineering in liquid (LDL) [2, 3]. In the following, these methods are briefly introduced.

LFL and LML were developed as techniques derived from LAL. These techniques use particle dispersion in liquids as raw materials and may also be classified as laser processing in liquids or laser processing of colloids (LPC). The goal is to shrink the colloidal particles, whereas the LML goal is to increase the scattered particles' size or deformity through colloids' laser radiation. LFL refers to the fragmentation of micro- and nanometer particles into smaller particles up to 2 nm [4]. The fragmentation is caused by the absorption of laser light and the evaporation of the photothermal or the

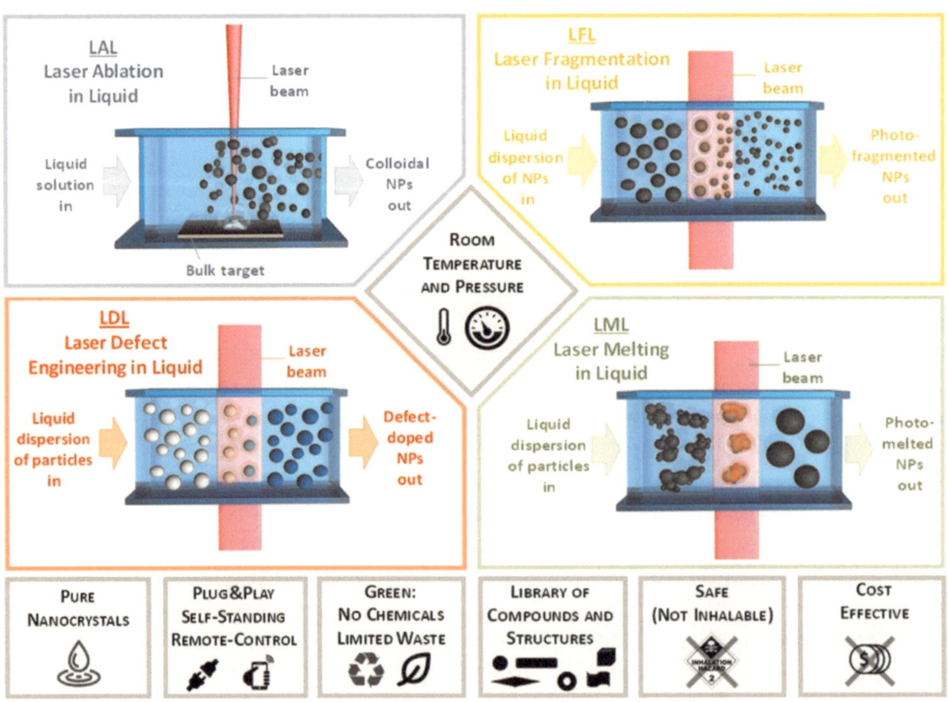

Figure 1.
The schematic of laser synthesis classification of colloids into LAL, LFL, LDL, and LML methods [2].

Coulomb explosion with the size of the specific particles. In contrast, the LML refers to the melting and intertwined agglomerated nanoparticles in the sub-micrometer and micrometer spheres [5].

The use of controlled laser fluence (poor laser fluence) of the laser beam to radiation to colloid aiming to change the atomic structure of nanoparticles by creating defects without changing the size of nanoparticles is called LDL. Like the LFL and LML methods, the LDL method irradiates scattered particles with lower laser fluence. It does not intend to shrink particles like LFL or melt particles like LML. Therefore, LDL summarizes all studies that aim to change the properties and density without altering the particle size. In LDL, defects not existing in the primary particles must be inserted into the final product [2].

This study focuses mainly on PLAL, which developed in the 1990s. This method has recently attracted increasing attention. Nanoparticles produced by this method make promising applications in the fields of catalysts [6, 7], optics [8, 9], biomedical [10, 11], polymer composites [12, 13], thin layers [14], and additive manufacturing [15–17]. It has been proven that this method can be economical and scalable at a multi-gram production rate [18, 19]. In addition, it allows nanomaterials to be produced from unstable phases. However, conventional chemical methods cannot obtain non-stable phases [20, 21].

Given these benefits, research into nanoparticle production using the PLAL method has been highly regarded in the past 10 years. The number of articles cited from 2008 to 2020 is presented in **Figure 2(a)**. Also, many of the periodic table elements specified in **Figure 2(b)** are produced and studied by the PLAL method [22].

Figure 3 shows the most common applications of the PLAL method in biomedical, catalysts, and sensors. Water, soil, and atmospheric pollution caused by human activity's spreading of toxic chemicals is becoming a serious problem worldwide. Given the significant advances in nanotechnology, there is an urgent need to develop

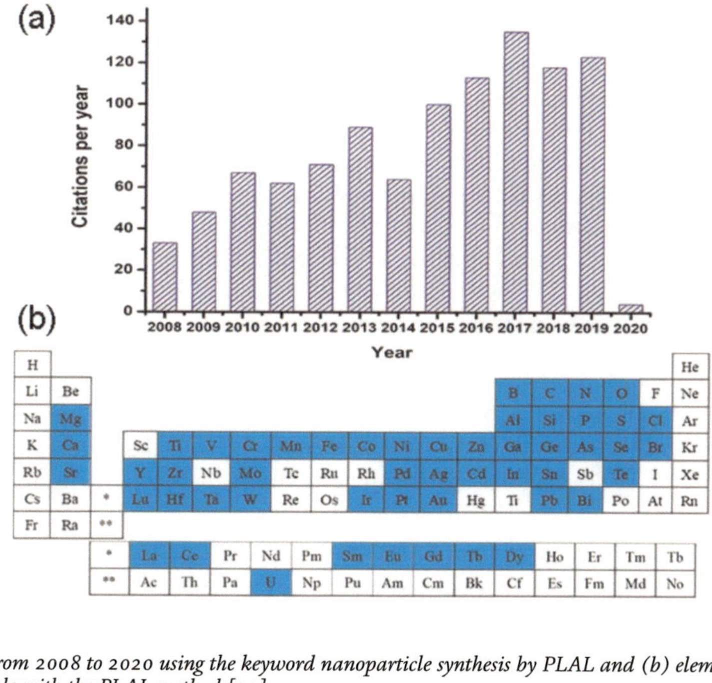

Figure 2.
(a) Citations from 2008 to 2020 using the keyword nanoparticle synthesis by PLAL and (b) elements specified in the periodic table with the PLAL method [22].

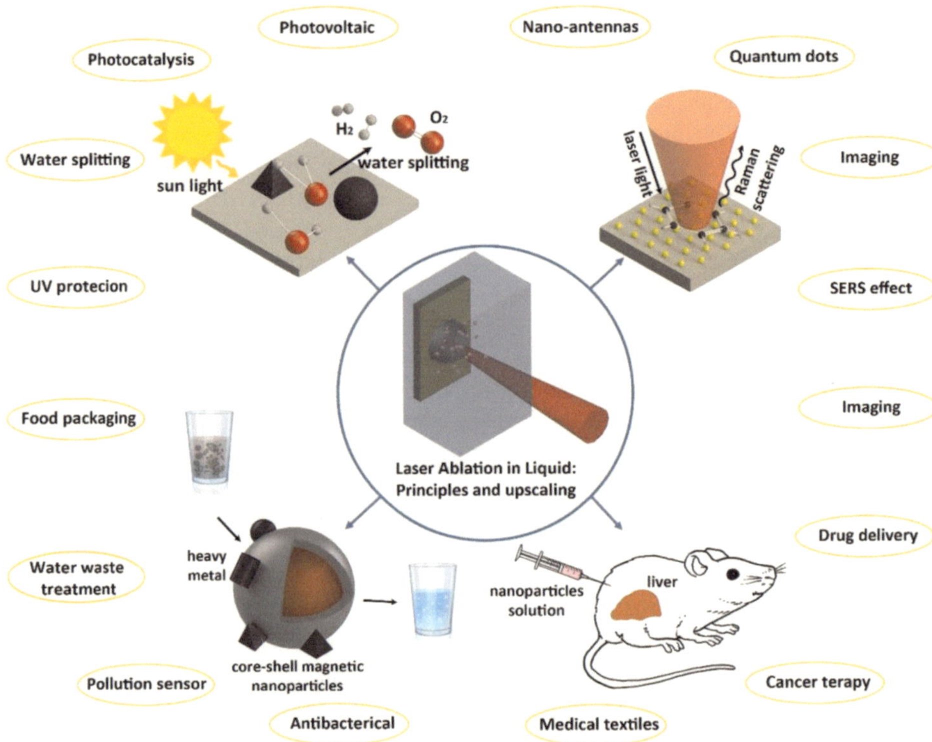

Figure 3.
Four important areas for PLAL [23].

green and economical approaches to innovative biomedical goals without environmental issues. Nanomaterials made with PLAL can successfully be used in biomedical, sewage treatment systems, and energy sources (e.g., solar cells and hydrogen production systems), and for producing clean and sustainable materials [23].

Therefore, the PLAL method as an alternative synthesis path for nanoparticle production in liquids has proven its ability to produce high-purity colloidal nanoparticles for various materials. The basis of synthesis with this method is using a pulsed laser. The PLAL method can create complex structures, such as core-shell structures, which are difficult to synthesize by typical chemical synthesis [3]. Also, researchers have increased the efficiency and reproducibility of colloidal nanoparticle production on a liter using a strong laser system [18, 19, 24]. Given the benefits, companies such as GmbH and Strem in the United States produce colloidal nanoparticles using the PLAL method [25, 26]. **Figure 4** compares the production efficiency of colloidal nanoparticles in PLAL and the chemical method performed by Jendrzej et al. [3, 19]. Accordingly, if gold nanoparticle production exceeds 550 $mg.h^{-1}$, PLAL synthesis is more economical than chemical synthesis.

3. Synthesis mechanism of nanoparticle by PLAL

The PLAL mechanism contains several physical processes. The main steps of PLAL are shown in **Figure 5**. This process begins with the absorption of the laser pulse by

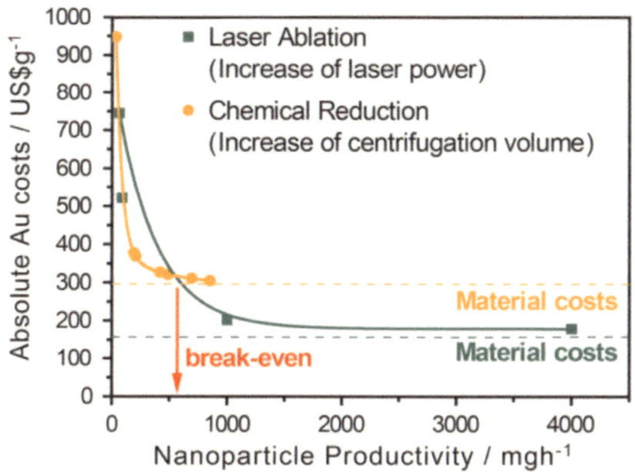

Figure 4.
Comparison of the costs of producing gold colloidal nanoparticles by PLAL and chemical methods [19].

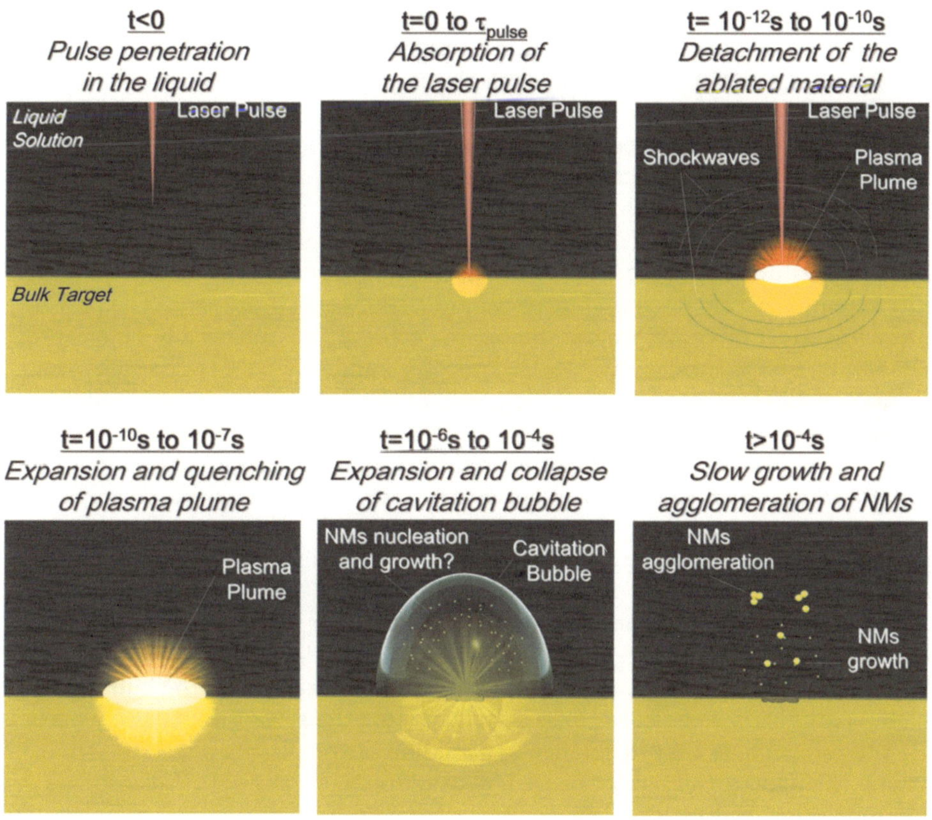

Figure 5.
The schematic of the main steps in the PLAL method [27].

the target. Then, a plasma plume containing the ablated material expands into the surrounding liquid and is accompanied by a shock wave. This expansion is due to the high pressure and temperature inside the plume. The plasma plume cools during expansion and releases energy into the liquid. This phenomenon creates a bubble caused by cavitation that expands in the liquid. Then, this bubble, due to cavitation, collapses on a time scale of hundreds of μs with the release of the second shock wave [27].

During the PLAL process, the metal target absorbs the laser photon energy and creates heating and ionization in the area under radiation. Laser energy stimulates the interatomic bonds of the metal target, causing them to break at the threshold energy level. According to bremsstrahlung, free electrons absorb the input laser photons and induce more ionization in the target matter. In addition, melting, evaporation, and plasma processes co-occur. These interactions cause the material to be removed from the solid target surface through vapors, liquid droplets, and solid fragmentation. The amount of ablated area depends on the absorbed energy (E). PLAL process parameters to estimate the ablated amount by the relation of $E^{1/3} \propto \tau_e$, $E^{1/2} \propto \tau_a$, $E^{2/3} \propto L_a$, and $\tau_a \gg \tau_1$ are related to each other, where in this relation, τ_1, L_a, τ_a, and τ_e are the laser pulse time, ablation depth, ablation process time, and the electronic temperature during the ablation process, respectively [28].

The atomization and ionization process produces dense, energetic, non-equilibrium plasma at supersonic speed. Because of its rapid expansion, plasma acts as a piston against the surroundings (liquid and target). It creates shock waves moving toward solid and liquid target in opposite directions. This shock wave increases the temperature and pressure inside the plume [28].

Giacomo et al. [29] showed that the shock wave created by the NS pulse may last several hundred microseconds in the water and up to a few millimeters before the collapse during cavitation. These researchers found that the external shock wave (toward the liquid environment) did not play an essential role in cavitation.

The plasma plume cools during expansion and releases energy into the liquid environment. The plasma plume turns off after 10^{-8} to 10^{-7} seconds. The process forms a thin layer of vapor around the plasma volume and creates a bubble caused by cavitation on a time scale of 10^{-7} to 10^{-6} seconds, and this bubble disappears on a time scale of 10^{-4} seconds. The bubble grows in the liquid to a maximum diameter of millimeter-scale [28].

During the movement of bubbles caused by cavitation, its internal temperature and pressure are reduced to the surrounding liquid. Then, the bubble disappears and releases energy by publishing a shock wave, affecting the phase transfer and nano-particle accumulation. After the bubble collapses on a time scale of 10^{-4} seconds, the system reaches a stable state physically and chemically. The accumulation of unstable particles is also possible depending on the particles' composition and surface oxidation. This bubble caused by cavitation acts as a rule for the nucleation of nanoparticles, their growth, and accumulation. The interaction of the bubble with the enclosed particles is an essential step in determining the size of the primary particles [28]. The following is a more detailed explanation of the plasma phase and the dynamic behavior of the bubble caused by cavitation.

3.1 Plasma phase

For example, **Figure 6** shows the temperature map of an induced plasma on a water-immersed aluminum target. By examining the figure, it can be seen that the plasma kernel has a temperature higher than the dense temperature of most metals. However, due to the high pressure of the plasma under water-enclosed conditions,

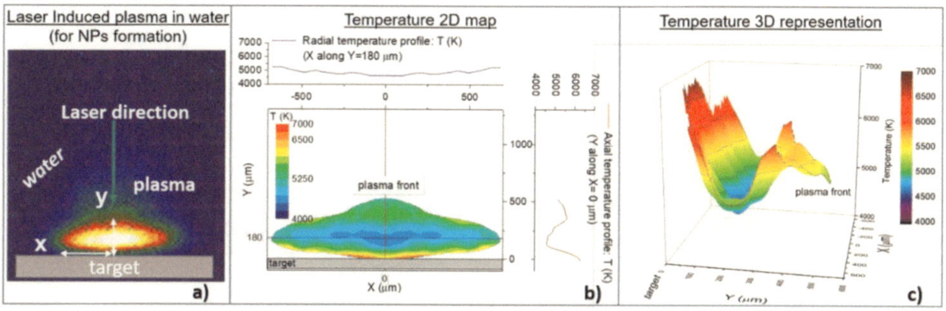

Figure 6.
(a) Image of laser-induced plasma on an Al target immersed in water taken with ICCD camera, (b) 2D map of calculated plasma temperature (laser energy = 270 MJ, laser beam diameter = 1 ± 0.2 mm, laser wavelength = 532 nm, laser pulse time = 6 ns, laser frequency = 10 Hz), and (c) 3D display of plasma temperature on the 2D map of part (b) [23].

it allows the density at higher temperatures compared to standard conditions [30, 31]. These observations show that nanoparticles can be formed at the plasma boundary (the area between plasma and liquid) and in the significant part of the plasma. In this case, particles of different sizes and shapes are predicted to be produced at the plasma boundary because the plasma boundary is out of equilibrium. In contrast, the particles formed in the major part of the plasma, in which the processes of growth and evaporation are in thermodynamic equilibrium, are obtained with a spherical form of beam size distribution [23].

The mechanism of growth in the plasma phase during the nanosecond (ns)-PLAL laser ablation process has been studied with a theoretical model in which the competition between evaporation and electrostatic growth is examined [31]. These observations show that the electrons are connected as soon as small clusters form due to the high concentration efficiency at the early ablation stage. As a result, the particles are negative and absorb the plasma ions. With the reaction of the ions with a negative charge, particles are formed (electrostatic growth), and particles begin to grow until the growth rate is balanced by the evaporation process due to the high temperature in the plasma (4000–6000 K). In this case, the equilibrium between electrostatic growth and thermodynamic evaporation regulates the dimensions of nanoparticles, and a spherical morphology is achieved. It has recently been shown that when the particle exits the plasma, it still has extra electrons. This additional negative charge can prevent the accumulation of nanoparticles due to repulsion [23, 32].

Electron density is one of the essential parameters to determine plasma properties. The electron density equals the number of free electrons in the plasma. Given the quasi-neutral state, the number of free electrons equals the number of charged heavy particles (ions) ($n = n^+ = n_e$, where n represents plasma density), in which case the number of charged heavy particles can be estimated. These charged particles may significantly affect the dynamics of chemical reactions through plasma chemistry. A general classification of PLAL-induced plasma compared to other plasmas such as solar corona, ionosphere, magnetron sputtering, glow discharge, flames, thermonuclear fusion, low-pressure arcs, high-pressure arcs, thermonuclear reactors, and magnetohydrodynamic energy conversion is shown in **Figure 7**. This figure shows that the plasma density in the PLAL is significantly higher than other plasma [33].

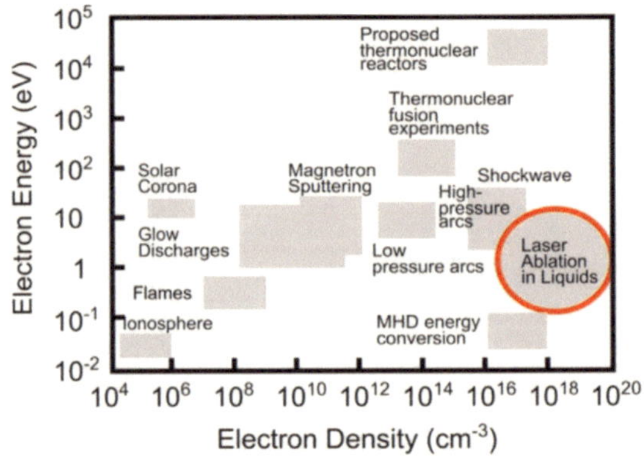

Figure 7.
The classification of PLAL-induced plasma compared to other methods [33].

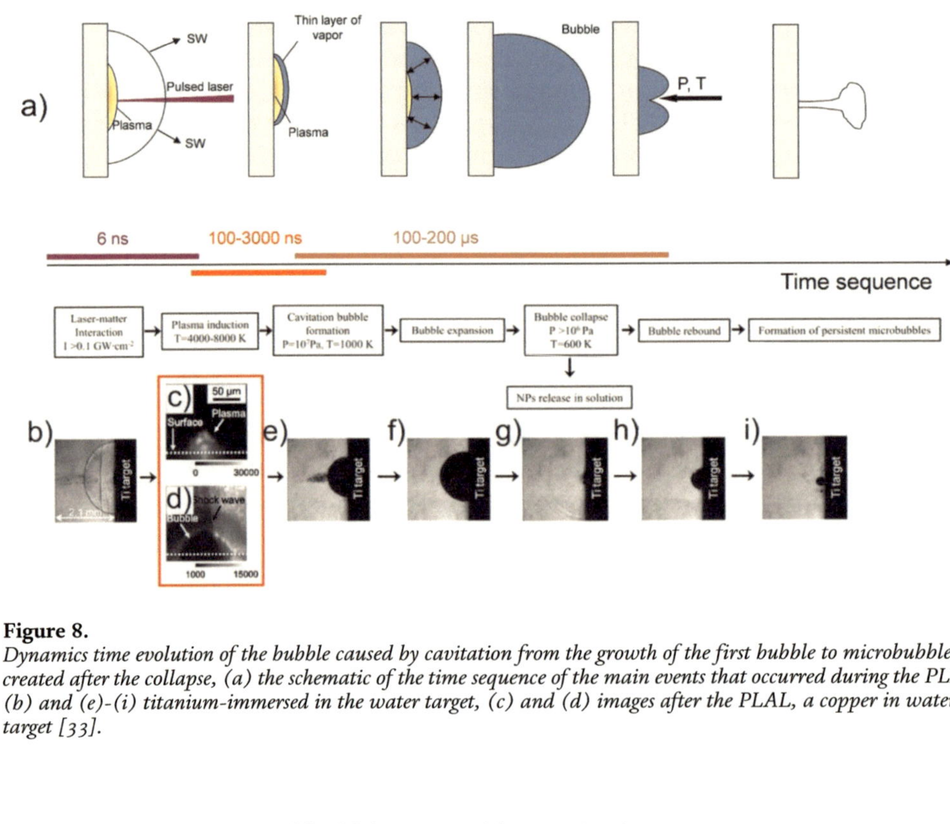

Figure 8.
Dynamics time evolution of the bubble caused by cavitation from the growth of the first bubble to microbubbles created after the collapse, (a) the schematic of the time sequence of the main events that occurred during the PLAL, (b) and (e)-(i) titanium-immersed in the water target, (c) and (d) images after the PLAL, a copper in water target [33].

3.2 Dynamic behavior of bubbles caused by cavitation

According to **Figure 8**, the development of bubbles caused by cavitation can be classified into three stages: bubble expansion caused by cavitation, contraction, and collapse (**Figure 8(a)**). The schematic shows the time sequence of the main events during the PLAL. **Figure 8(b)** and **(e)-(i)** show images of the time evolution of the bubbles caused by cavitation after the interaction of a laser pulse with the aim of water-immersed titanium. **Figure 8(c)** and **(d)** show images of PLAL of a

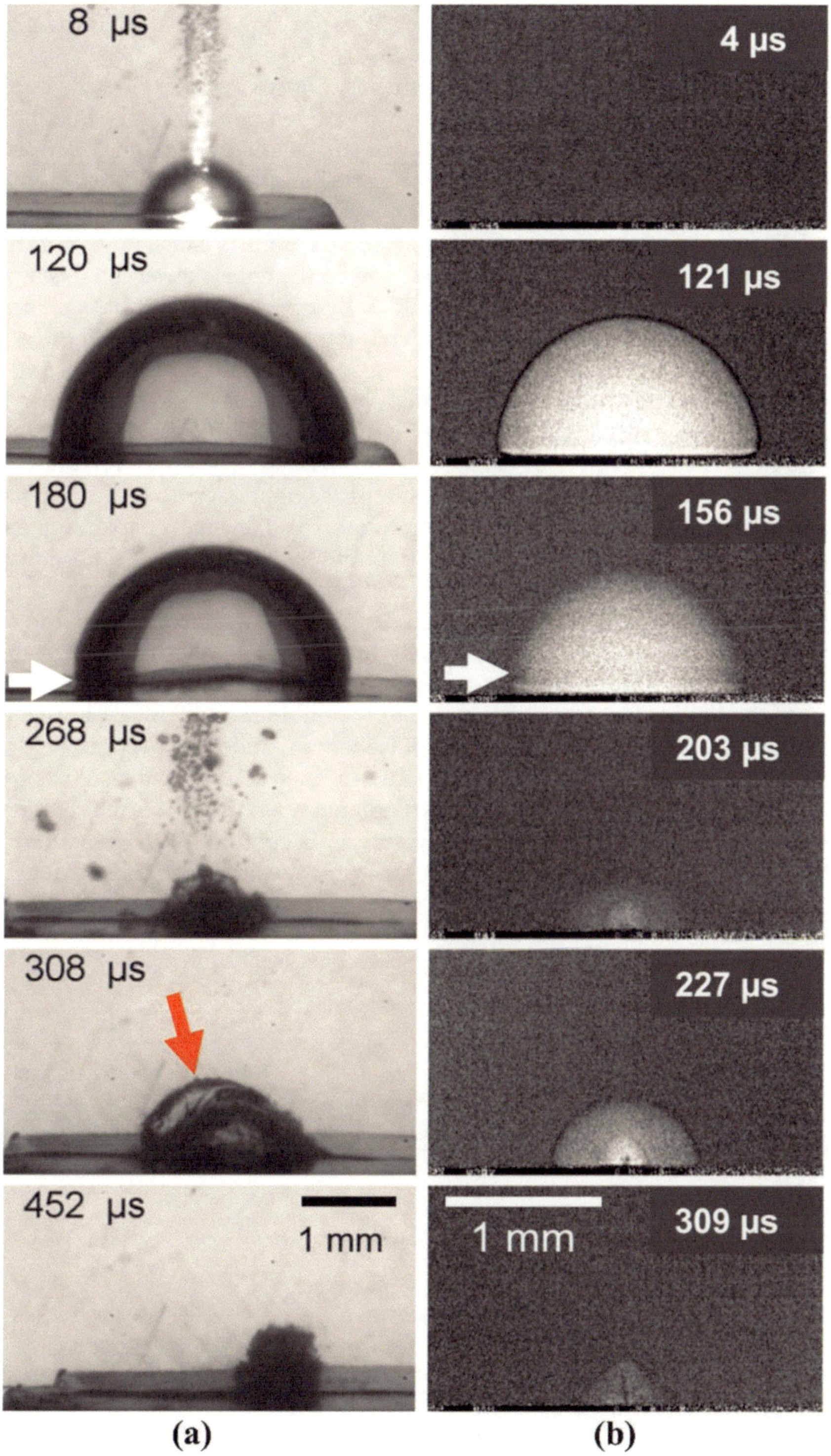

Figure 9.
(a) Stroboscopic images and (b) X-ray radiographic imaging of the PLAL, silver-immersed in the water target (the white arrow indicates the formation of the bubble wall with a pseudo-hemisphere and the red arrow shows the depression and deformation of the bubble) [33].

copper target in the water where a thin layer of vapor is formed around the plasma boundary [33].

In stroboscopy and X-ray radiography images, **Figure 9** shows that the bubble caused by cavitation expands and, after reaching a fixed point with the maximum size and shape of the pseudo-hemisphere, begins to shrink and eventually collapse. For each collapse, part of the mechanical energy of the bubble caused by cavitation is released through the release of a new shock wave [33].

According to **Figure 10**, during the expansion of bubbles caused by cavitation, the formation of tiny microbubbles around the growing bubble is observed. Using single-pulse experiments, Tanabe et al. [34] concluded that these microbubbles are related to pre-existing nanoparticles. These nanoparticles act as targets by absorbing the laser beam and forming microbubbles.

Shih et al. [35] showed that tiny satellite bubbles around the bubbles caused by cavitation were also created with PS laser pulses when applying single pulses in liquids where the PLAL process has not been performed. They suggested that larger, faster-formed nanoparticles are placed at the bubble phase boundary caused by cavitation and are responsible for forming satellite microbubbles. However, explaining exactly how the presence of large nanoparticles at the bubble boundary due to cavitation causes the microbubbles still requires further studies [33].

Figure 11 shows that the collapse of bubbles caused by cavitation can lead to the formation of microbubbles that have a much longer life (in ms to s) and are about a few millimeters. The time of growth and the stay of these microbubbles depend on the viscosity of the liquid. The lifespan and size of the bubble caused by cavitation depend on laser parameters such as laser density, pulse length, and fluid properties [33].

Laser density is critical because the increase in laser density increases the bubble size caused by cavitation and the amount of ablated mass. Based on the relation (1), the maximum bubble radius at high energy density is obtained. In this relation, Rmax, ρ, tcollapse, Pin, and Pv are the maximum radius of the bubble, the fluid density, collapse time, the liquid pressure, and the vapor saturation pressure inside the bubble, respectively [28].

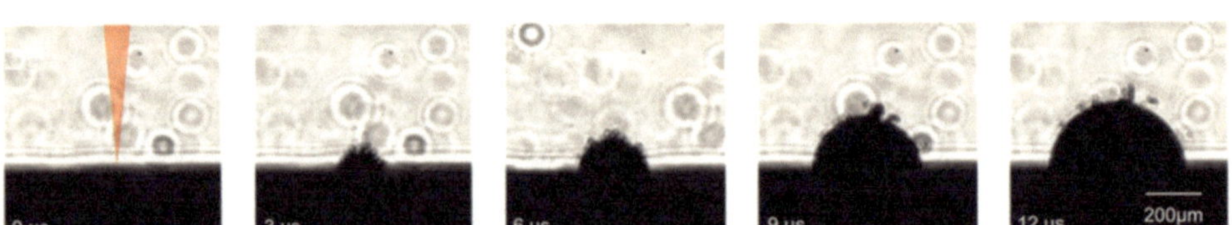

Figure 10.
The formation of satellite microbubbles around the bubble caused by cavitation after single-laser pulse radiation (2 ps) to a gold target [33].

Figure 11.
The formation of sustainable microbubbles of a few ms after the PLAL, a gold-in-water target [33].

$$Rmax = \frac{tcollapse}{0.915\sqrt{\rho/(Pinf - Pv)}} \tag{1}$$

Reich et al. [36] found a linear relation between bubble size caused by cavitation and higher ablation efficiency than the laser density threshold (39 J.cm^{-2}) for the ns-PLAL, which showed almost the entire laser energy on the laser density threshold (39 J.cm^{-2}) is used for the formation of bubbles (**Figure 12**).

4. Research methods in PLAL

The direct study of laser ablation in fluid is very challenging for the following reasons:

- Most structural study techniques do not have the time and space resolution required for the highly rapid dynamic monitoring of the PLAL process.
- The amount of material required to study is in the range of 10^{-1}–10^{-3} μg per pulse because the sensitivity of most of the techniques is structural study.
- The light emission of the plasma plume and the laser pulse interfere with most optical absorption spectroscopy techniques.
- To measure the size of the nanoparticles in the liquid, such as the analysis of Dynamic Light Scattering (DLS), which operates based on volume weight signals. This means that a micrometer material separated from the target produces more signals than thousands of nanometer nanoparticles, thus not measuring nanometer particles.

So far, many methods have been used to investigate the process of laser ablation and nanoparticles at various moments, presented in **Figure 13**. **Figure 14** also includes different characterization methods for different PLAL stages with time scales. The following is an explanation of characterization methods.

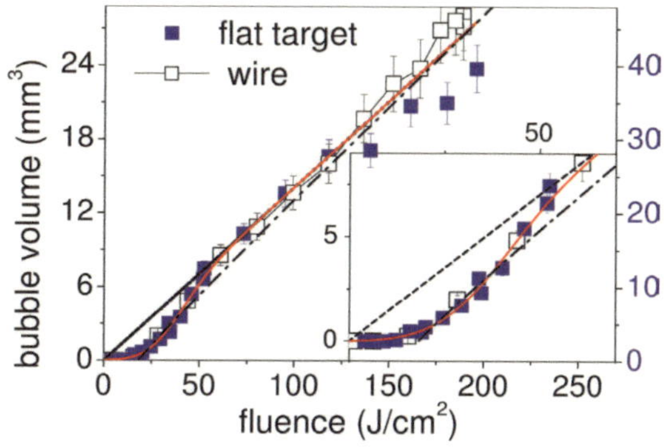

Figure 12.
The bubble volume produced by the PLAL, a flat and wired silver target, in terms of laser density [36].

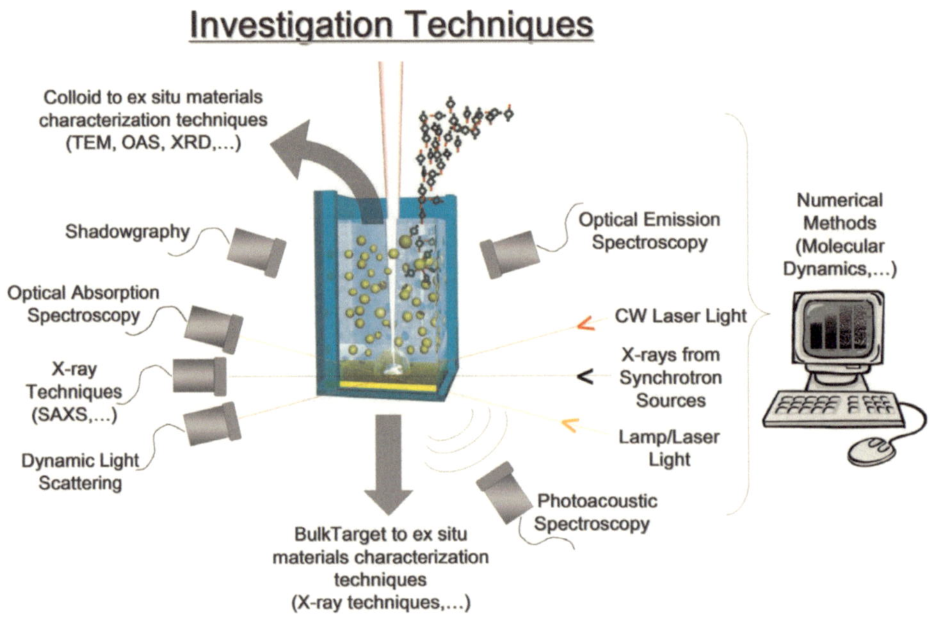

Figure 13.
The schematic of the methods used to study the PLAL method [27].

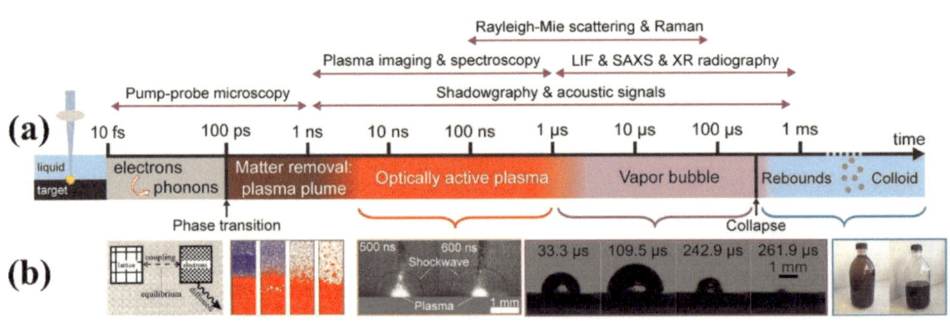

Figure 14.
(a) The different steps in the PLAL method for the ultrashort pulses and the characterization methods at different times and (b) from left to right, respectively: The schematic of electron-phonon, phase transition transmission simulation using molecular dynamics modeling, plasma and released shock wave, bubble dynamics, and colloid produced [2].

- The optical emission spectroscopy (OES) is performed with a time separation for plasma detection with a timely resolution below the ns and a sub-millimeter spatial resolution with the charged coupled device (CCD) camera.

- Ns shadowgraph imaging provides information about the time evolution of the plasma target, shock waves, and cavitation bubbles with micrometer spatial resolution.

- Photoacoustic spectroscopy can be performed at a temporal evolution of 10^{-6} seconds, which provides information about the time evolution of shock waves.

- Ultraviolet-visible spectroscopy (UV-VIS) can be performed with a CCD camera with a millimeter space resolution.

- DLS provides information about the presence of objects of a few nanometers or larger with temporal evolution of minute. This technique does not distinguish between bubbles and nanoparticles; larger objects heavily influence signals.

- Recently, X-ray techniques have been used to investigate the real formation time of nanoparticles during laser ablation. However, these methods require high-temporal resolution X-ray sources such as synchrotron radiation to obtain detectable signals with ns or ms temporal resolution. Different X-ray techniques, such as X-ray diffraction (XRD), small-angle X-ray scattering (SAXS), and X-ray absorption spectroscopy (XAS), have synchronous sources. Among these methods, they are mainly focused on SAXS. This is probably because it gives relatively intense signals and provides information about the size of the nanoparticles rather than their composition.

In general, to remove existing limitations, theoretical and numerical models can ideally overcome most of the limitations of empirical research methods. Laser ablation and nanoparticle formation can be modeled by molecular dynamics with an atomic spatial and almost unlimited temporal resolution. Molecular dynamics calculations consider the non-thermodynamic equilibrium of laser ablation, and they usually examine electron-ion collisions using Monte Carlo's method. One of the drawbacks of molecular dynamics measures is that they do not consider ultrafast nonlinear light phenomena, such as photoionization or multi-photon absorption, for charge separation fields from space and Coulomb explosions [27].

5. Effect of the PLAL process parameters on the production of nanoparticles

The PLAL process parameters are adjustable and related to the process components, including the type of laser, the laser optical adjustment components, the ablation chamber, the target piece, and the liquid environment. A process begins with the input of the primary parameters and creates an output that includes the results of the process. Variables can be defined between these parameters and the process results. The dependency graph in **Figure 15** shows a combination of PLAL process parameters. In addition, the dependence of the parameters on each other in the process variables and, finally, the result of the process (product) is also shown in the figure. For example, the laser pulse width parameter directly affects the output efficiency and indirectly affects the laser current intensity variable. Pulse width colors (green) and laser fluence (purple) reach a dark brown, indicating an existing intensity variable. It should be noted that product characteristics such as particle size distribution also represent an output of the process but are not considered in detail here [23].

5.1 Effect of laser wavelength

Laser wavelengths affect the amount of ablation in the PLAL. Almost all studies show higher efficiency using IR laser light than UV or VIS. However, in the PC

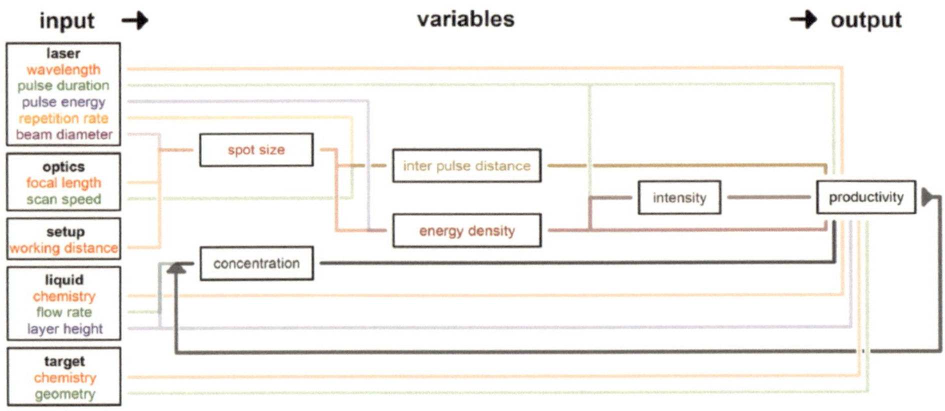

Figure 15.
The dependency graph of the influential and interdependent process parameters in the PLAL process [23].

process, nanoparticle protection due to Rayleigh scattering [37] can also cause differences in efficiency.

In addition, for almost all metals, higher absorption cross sections are presented for nanoparticles at UV or VIS wavelengths compared to IR wavelengths [38].

The short laser wavelength increases the metal target's absorption energy and the efficiency of the nanoparticles produced. However, this effect is insignificant due to its open absorption effect that increases ablation efficiency, especially in noble metal materials such as silver, gold, and platinum. This is due to the exacerbation properties of the plasmon in the UV-Vis area. Due to this property of synthesized nanoparticles in the liquid environment, they will be able to re-absorb the input laser pulse, which has adverse effects, reduces the ablation rate, and expands the size distribution of particles. The produced plasma plume can also re-absorb the laser with a shorter input wavelength. However, a near-infrared laser wavelength can eliminate this open absorption effect. The distinct absorption of target matter at different wavelengths results in the production of nanoparticles at different concentrations that affect the size of the particles. However, it should be noted that nanoparticle morphology does not change [39].

It can be concluded that higher ablation rates (higher efficiency) for metals generally occur throughout the UV laser wavelength due to inter-band absorption. However, the efficiency in VIS and IR wavelengths can be comparable to UV wavelengths. There is also severe dependence on the metal target, laser fluence, and laser pulse duration. In the case of high colloidal concentrations, the absorption of shorter wavelengths by particles can reduce ablation. These relationships are summarized in **Figure 16**. It is worth noting that for semiconductor and dielectric materials, the initial ablation rate in the absence of a light-extinction colloid can be higher at red and near-IR wavelengths under certain conditions due to the difference in the ablation mechanism compared to metals [23, 40, 41].

Also, when laser ablation is performed in a liquid environment, the effect of wavelength is very noticeable because, as mentioned earlier, the particles are placed in the path of subsequent laser pulses. On the other hand, the nanoparticles produced depend on the composition of the target matter and have different absorption at different wavelengths. For example, the silver nanoparticles absorb 0.1, 0.28, and 0.52% of the 1064, 532, and 352 nm wavelengths, respectively. Therefore, the self-absorption effects of colloids are more significant at shorter wavelengths. Sending a

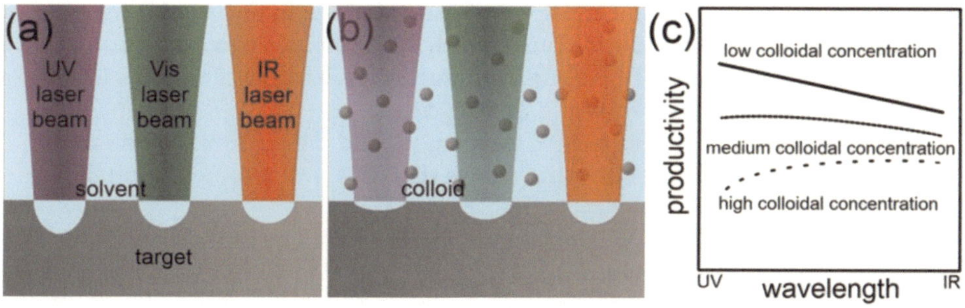

Figure 16.
The schematic of laser ablation of metals in liquids in equal laser fluence in different wavelengths, (a) in the absence of colloidal particles, (b) in the presence of colloidal particles, and (c) the effect of increased colloidal concentration on nanoscale efficiency in the PLAL process in different wavelengths [23].

wavelength at which the absorption of nanoparticles is high causes thermal stimulation, and as a result, the particles are crushed. It should be noted that when self-absorption is small, the size of the particles is determined by the penetration depth of the laser light in the target [42].

Solati et al. [43] reported that radiation of laser pulses at 1064 nm reduced the size of Ag colloids because of the self-absorption of laser pulses by colloidal particles. Tsuji et al. [44] reported that the increasing radiation of colloids by laser after laser ablation results in smaller particles. This phenomenon was also observed in the case of gold colloids prepared using 1064- and 532-nm lasers. Amendola and Meneghetti [45] also examined the laser ablation of gold nanoparticles at 532 and 355 nm. They reported that these wavelengths led to the expansion of the size distribution and a decrease in the average size of nanoparticles. In addition, laser-induced nanoparticles can increase the reaction to other species in the solution and destruction of adjacent organic molecules.

These values have shown that the size of the colloids can be controlled by changing the total number of laser pulses. Also, the size of nanoparticles can be altered by changing the energy of the laser photon (wavelength). Still, it should be noted that an increase in the wavelength will lead to increased particle size [46].

5.2 Effect of laser beam radiation time

The observed absorption features of the UV-Vis spectrum showed that the size of the synthesized nanoparticles is related to laser ablation time (LAT) and laser fluence (LF). A set of experiments determined the effect of parameters on nanoparticle size. Increasing ablation time increases nanoparticle density and decreases liquid molecules [47].

The number of radiation pulses is also adequate in the production of nanoparticles. At the same time, the laser wavelength is 1064 nm, and the total number of pulses is low; the production of nanoparticles is increased linearly with the increase in pulses, but when the laser wavelength is 532 nm. The total number of laser radiation pulses is very high; production is much lower than the pulse number. Initially, the production of nanoparticles increases linearly with the increase in the number of pulses. However, when it reaches saturation, it is due to nanoparticle absorption of laser light [42]. To prevent saturation, the liquid must enter the container from one side and exit from the other, in which case the concentration of nanoparticles in the colloids increases for many laser pulses [27]. An example is the synthesis of copper nanoparticles in a laser fluence of 151 $J.cm^{-2}$ and LAT of 10 to 50 minutes by Desarkar et al. [48]. Mie theory was used to investigate the results and extract data.

Figure 17(a) shows changes to SPR wavelength and particle size (diameter in nanometer) regarding LAT. This figure shows the TEM measurements and SPR changes with LAT.

As shown in **Figure 17(b)**, it is observed that the width of the surface plasmon resonance line (bandwidth) is inversely proportional to particle size. In the mentioned figure, a Lorentzian proportion was used to determine the width of the Lorentzian line (according to meV) of the SPR peaks. As the ablation time increases, the volume of nanoparticle production is initially increased and then fixed. The target nanoparticles protect the piece from laser radiation, which is why, after reaching the critical time, no new nanoparticles are produced, and the extra radiation of laser pulses on the colloid through the laser ablation of the liquid phase reduces the size of the nanoparticles which this mechanism (fragmentation of nanoparticles due to the self-absorption phenomenon) is as shown in **Figure 18** [49].

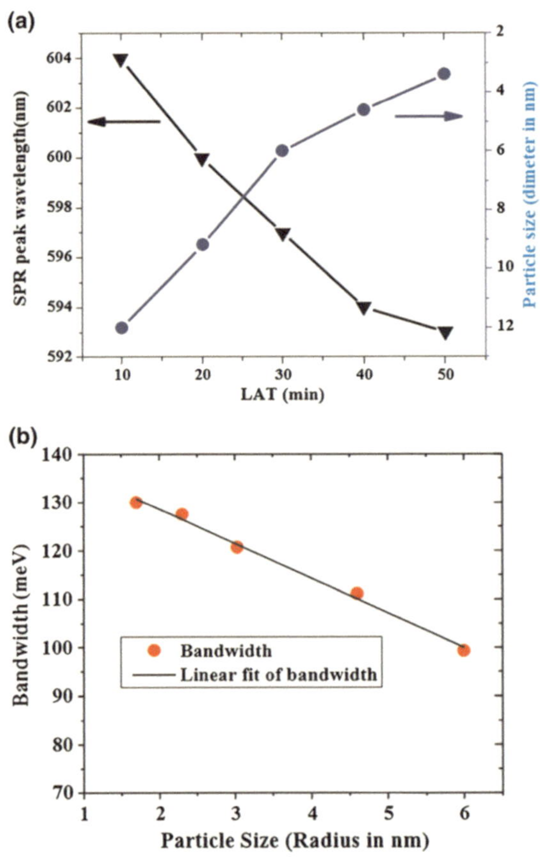

Figure 17.
(a) Changes of the SPR wavelength and the average particle size (diameter in nanometer) obtained from the analysis of TEM images with the LAT (min) and (b) line width change (meV) with particle size (radius in nanometer) [48].

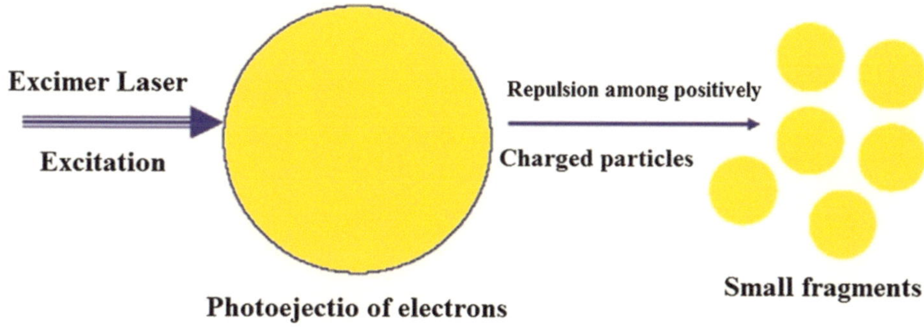

Figure 18.
The process of nanoparticle fragmentation due to excess laser radiation [49].

5.3 Effect of laser pulse time changes on the size of nanoparticles

The pulse time (pulse width) is an essential parameter in the synthesis of nanoparticles. Laser pulse width is effective in the production and size of nanoparticles produced by the laser ablation method. There are two theories based on the size of nanoparticles and the efficiency of nanoparticle production at different pulse widths. In the following, these two theories will be explained along with their analysis.

The first theory reported that the size and efficiency of the nanoparticles increase during the longer pulse time (ns) over the shorter pulse time (fs and ps). The efficiency of nanoparticle production and size is believed to increase with the depth of the molten layer on the target surface. The cavity produced by the fs laser pulses is much shallower than the cavity made by nanocrystal laser pulses, indicating lower efficiency of nanoparticle production. For this reason, the efficiency of ablation and size of nanoparticles produced with ns laser pulses is greater than that of fs laser pulses. Also, in ns laser ablation in the liquid environment, plasma is created after the laser pulse; therefore, the plasma's life is very low. For this reason, the amount of nanoparticle production decreases. Reports show that plasma pressure plays a vital role in the amount of material being slowed down from the target surface in laser ablation.

In the second theory, it is believed that by changing the pulse time from ps and fs to ns, the size of the particle will be reduced, and efficiency will decrease. It is suggested that by changing the pulse time from the ns to the ps and the fs, the ablation mechanism changes from melting and thermal evaporation to phase explosion. The shorter pulse time leads to a more efficient ablation process, resulting in instant evaporation and a minimum heat-affected area. In ps, ablation is faster than ns due to the lower threshold limit for metals [50]. It has also been reported that the energy absorbed by the target for ultra-short laser pulses remains very low in the target piece. As a result, extra-short laser pulse time (ps and fs) is beneficial because the heat loss is low, and the ablation efficiency increases [51]. The thermal nature of the laser ablation is intensified over a longer pulse. Heat losses become more prevalent in this case, so the ablation rate decreases. In addition, the life of the plasma-induced with laser and pulse time on a comparable time scale (more than ten or one hundred ns) for laser pulses intensifies self-induced plasma protection from laser pulses.

Figure 19 shows the remaining action point by colliding a laser pulse with an fs, ps, and ns laser. The remaining point of action by the laser pulse on the target clearly shows the occurrence of thermal ablation processes in the gas and liquid phases. In the

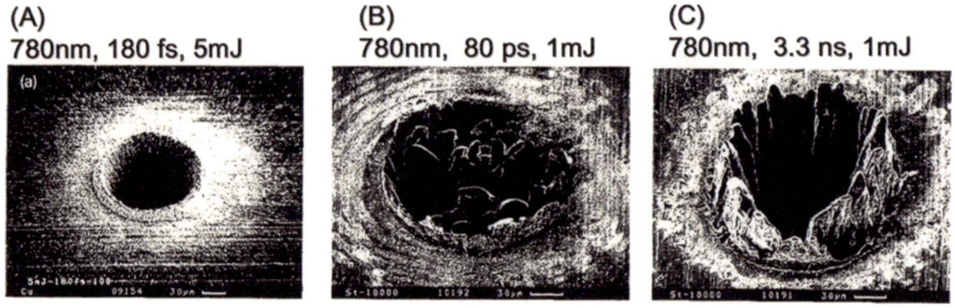

Figure 19.
The remaining effect of the action on metal targets after laser ablation with pulses (a) fs, (b) ps, and (c) ns [52].

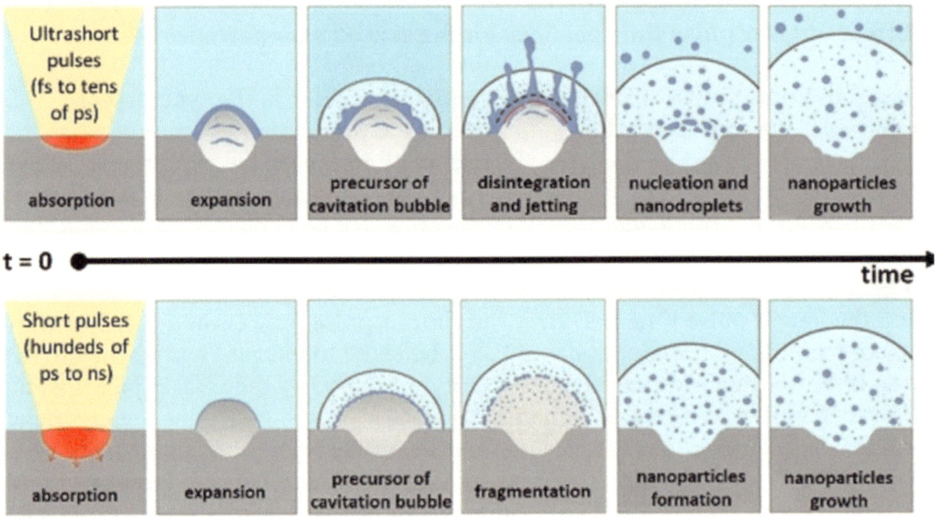

Figure 20.
The schematic of the PLAL process with lasers with ultra-short pulses (fs to dozens of ps) and short (hundreds of ps to ns) [2].

fs state shown in **Figure 19(a)**, the remaining point of action has sharp borders and corresponds to the laser point of action (diameter size of the laser beam), an example of ultra-fast local heating. In this case, the process is mainly influenced by direct photoionization. In the ps pulses (**Figure 19(b)**), the action point is less than the fs state (**Figure 19(a)**). It has sharp borders, an example of the simultaneous direct photoionization and thermal ablation processes (melting and evaporation) that have been driven out due to the high pressure created by the material [27].

With ns pulses, ablated materials and laser pulses will be used together due to the plasma protective effect for a relatively long period. This time is sufficient to transfer part of the laser energy to the plasma plume and thus increase its temperature and pressure. In such conditions, as shown in **Figure 20** as a schematic, the melted droplets from the target into the plume have a better chance of maximizing evaporation [2]. This mechanism also prevails in the ns state for the shrinking nanoparticles. Researchers have reported this mechanism for gold laser ablation, and the size of nanoparticles obtained in fs and ps pulses is larger than in ns pulses [45, 53, 54]. In addition, it has often been observed that the size distribution of nanoparticles

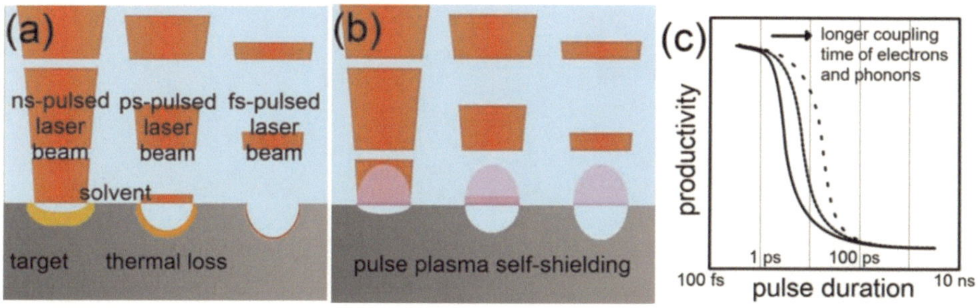

Figure 21.
An image of laser ablation of metals in liquids during different pulse times for (a) heat losses, (b) self-protection by laser plasma during different pulse times, and (c) effect of further coupling of electrons and phonons on nanoparticle efficiency produced in the PLAL process during different pulse time [23].

obtained with ns pulses is thinner than with ps and fs pulses. This may be another effect of improved homogeneity due to the overlap of the plasma plume with the laser pulse [27, 45].

As shown in **Figure 21**, in the fs laser pulse (10^{-15}), the laser energy is released to the electron in the metal target faster than the thermalization process of the electron-phonon target. In the ps laser (10^{-12}) and ns laser (10^{-9}), the thermal relaxation process is more remarkable, which leads to energy being released thermally to the liquid environment before the end of the pulse. In a few tenths of ps of the laser radiation, plasma is produced and takes tens of ns after ablation. Therefore, no time overlap between the ablated materials and the laser pulse does not occur in the ps and fs laser. However, in the ns laser pulse, there is an overlap in ablated materials due to heat conductivity [28]. The long laser pulse time leads to the absorption of the input laser energy in the plasma plume and increases the plasma temperature and pressure. The plasma then atomizes the materials in the plume. This process homogeneously makes the ablated materials. The energy absorbed by the metal target is reduced because the plasma plume creates an optical protector around the metal target [45].

In μs and ms pulses, the mechanisms of thermal ablation are largely dominant because the material is separated by melting and evaporation, while the formation of a plasma plume is no longer required for laser ablation. The nanoparticles and atoms evaporated from the target are thrown into the liquid in which three types of reactions include: (a) the reactions of the target vapor phase and liquid solution at temperatures and pressure less than the plasma plume, (b) the reactions of the target liquid phase and the liquid solution, and (c) the reactions of the solid phase from the target and the liquid solution, may occur [27].

Figure 22 also offers a schematic of the laser ablation process at different time scales for fs and ns laser pulses. Laser energy transfer to the target, target phase transfer, plasma life, and bubble life occur on the order of multiple ps, above 100 ps, several μs, and several hundred μs, respectively [33].

5.4 Effect of the repetition rate of laser pulses (RRLP)

The distance between the two continuous laser pulses determines the quantity called the pulse repetition rate (frequency); therefore, increasing the pulse repetition rate will mean reducing the time between pulses. The pulse repetition rate can change the average size of nanoparticles, which is usually a nonlinear dynamic process in laser

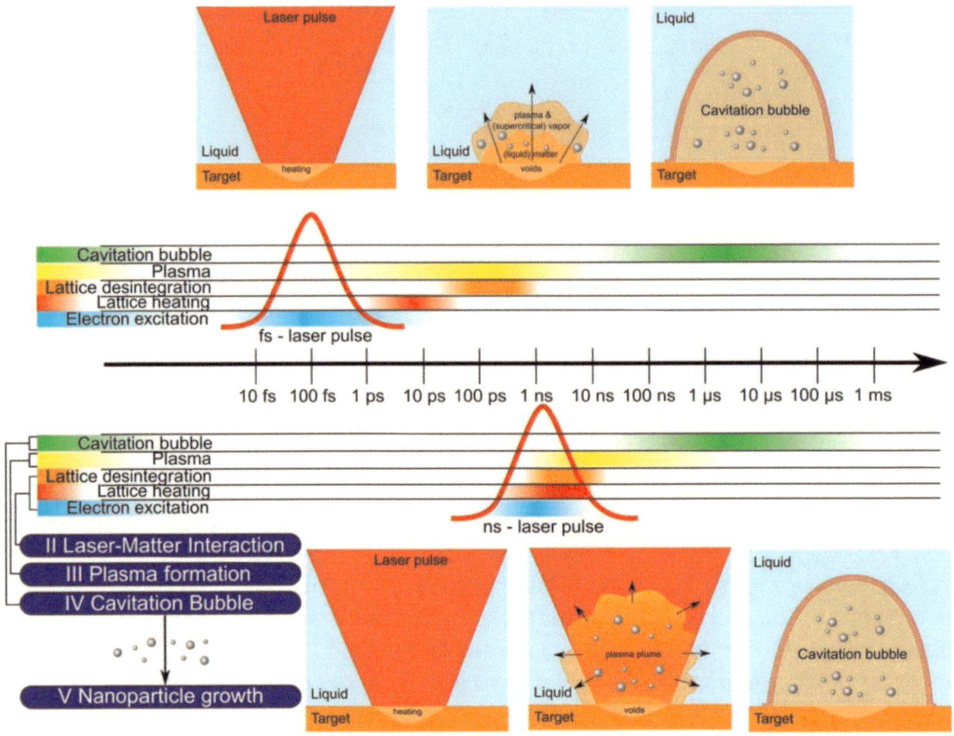

Figure 22.
The schematic of the laser ablation process at different time scales [33].

ablation. Initially, nanoparticles are produced by a laser pulse. During the subsequent pulse collision, the temperature drops sharply, and this time is an excellent opportunity for particles to cool down and stick to other particles. This process increases the size of the particles, and as the next pulse arrives, the particles are crushed by the absorption of this pulse and converted to smaller sizes. Therefore, in the process of laser ablation, targeting, and, on the other hand, fragmentation of the particles occurs by the above two mechanisms. Lumping and fragmentation interact with each other during ablation [23].

The longer the two-consecutive pulse distance, the more particles can bond with other particles and increase in size. As a result, when we use higher repetition rates, we expect the size of the particles to be smaller. On the other hand, the higher repetition rate (KHz) increases effective ablation due to the greater heat density it creates on the target [42]. Valverde et al. [55] have analyzed the effect of RRLP at 1 to 10 Hz on the synthesis of silver nanoparticles (Ag-NPS) by laser ablation in ethanol. The results showed that the efficiency of silver nanoparticles was decreased by reducing the RRLP from 10 to 1 Hz.

However, linear efficiency increases only at repetition rates exceeding the bubble's life caused by cavitation (10^{-3}–10^{-4} s). This is related to repetition rates below 10^3–10^4 Hz. The bubble caused by cavitation is characterized by the failure of the refractive coefficient at the liquid/gas interface, which disperses the laser light and reduces the laser energy to reach the target. In addition, during the expansion of bubbles caused by cavitation, target laser ablation is performed in a hot phase with low density, similar to laser ablation in gas. These effects are limiting factors for laser ablation at high laser pulse repetition rates on the target, which is the easiest solution to improve the efficiency of the laser ablation process. However, it has not yet been determined what other possible changes in the mechanism of nanoparticle formation when overlapping laser pulses with bubbles caused by cavitation can be.

Increasing the repetition rate also increases the local concentration of nanoparticles in the target/liquid interface because enough time to distribute nanoparticles in the liquid is reduced away from the ablation area in the liquid. This effect can lead to nanoparticle agglomeration. It also results in the dispersal of input laser pulses and thus reduces ablation efficiency [27]. Therefore, to achieve the distribution and size of smaller particles, it is necessary to optimize the RRLP because, at low RRLP, the particles can be intertwined due to enough time. At high RRLP, due to insufficient time to distribute in the liquid, the particles can be agglomerated.

5.5 Laser fluence effect

Laser fluence describes the pulse energy penetrating the effective area on the ablation target surface. In other words, the laser fluence is the energy of each pulse per unit area. The laser fluence threshold helps ablation efficiency. The ablation energy threshold refers to the minimum density (optical energy in each pulse area) needed to separate materials from the area under radiation in the target metal and the production of plasma [28]. Laser ablation in the liquid can synthesize nanoparticles in the different laser fluences. Pulse energy or working distance changes may realize laser fluence change. Different results by both methods can occur due to changes in the release of beams in the fluid phase.

By changing the intensity of the laser light, several parameters, such as target absorption and nanoparticle absorption, change simultaneously. Increasing laser fluence on the target surface increases the temperature and improves ablation. The laser's interaction with matter depends on the laser fluence. When the laser fluence increases, the maximum metal temperature will increase during the warming process. Therefore, despite the high density, the possibility of atoms arousing and ablation rises [56] and increases the efficiency of nanoparticle production [57]. The volumetric ablation rate increases logarithmic with laser fluence. The progress of the logarithmic function depends most on the mechanism of ablation, the surrounding liquid, and the target matter. If the efficiency in PLAL is more related to the laser pulse energy, the optimal laser fluence can be obtained in this case [23]. **Figure 23** shows the dependence of nanoparticle production efficiency on laser fluence and beam diameter.

When the working distance and diameter of the beam are constant, the amount of ablation and production of nanoparticles increases with increased laser power. However, in the case that power is considered consistent and the diameter of the beam is changed, as the beam diameter increases, the efficiency increases with the smaller beam diameter.

Higher pulse energy separates the material from the metal and increases the target metal concentration in the plume; therefore, with less energy, it is possible to distribute the thinner size of the nanoparticles in laser ablation. In addition, the higher laser fluence also extends the bubble's lifespan, leading to the bubble reaching a maximum radius [28].

An example is laser ablation of an aluminum target with laser Nd: YAG and a wavelength of 1064 nm for synthesizing Al_2O_3 nanoparticles. After the process of

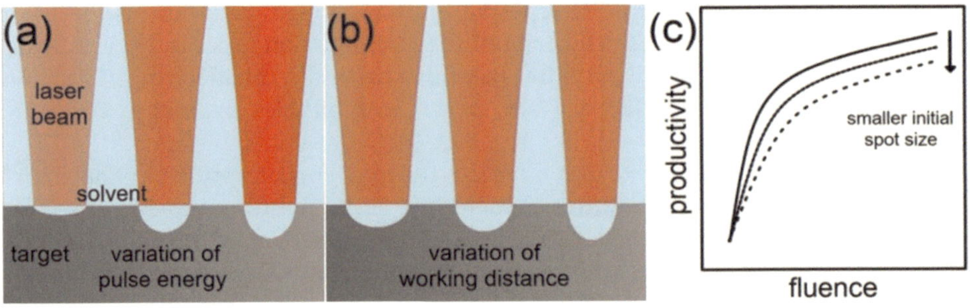

Figure 23.
An image of metal laser ablation in liquids in (a) different pulse energy (fixed beam diameter), (b) different working distances, and (c) indicating laser fluence changes in different sizes of beam diameter on nanoparticle production efficiency (Gaussian beam profile is considered) [23].

laser ablation, the liquid becomes opaque. As the laser fluence increases, the particle size varies from 27 to 49 nm. However, as the laser fluence increases, the self-absorption of nanoparticles will prevent the piece target ablation and lead to the crushing of the previous nanoparticles [58]. Mafune et al. [59] also studied gold laser ablation with ns laser and wavelength of 1064 nm in the energy range of 10–100 MJ. They observed that with the increase in pulse energy, the size distribution and the average size of nanoparticles increase.

On the other hand, the concentration of materials increases with increased energy. In other words, with increasing pulse energy, multiple mechanisms of separation of materials, such as fragmentation, phase explosion, and evaporation, co-occur. For example, high-energy gold laser ablation has distributed 2D size due to the simultaneous ablation mechanisms. In other studies, the distribution of single-dimensional size is obtained in low energy [54, 60]. Researchers have reported similar laser ablation results for platinum nanoparticles [61].

The theory of nucleation and growth can be used to explain the difference in particle size. Low laser fluence has less nucleation, leading to a smaller nucleus and particle size. In contrast, in high laser fluence, more nucleation results in large amounts of the nucleus and larger particle size [62].

5.6 Effect of laser focal length on nanoparticles produced

Concentration conditions (target position with focal point) are critical to forming a beam size distribution of nanoparticles. By changing the distance between the lens and the target, the diameter of the laser beam changes. The focal length affects the distribution and size of nanoparticles [63]. It has been observed that the result of ablation is highly dependent on local conditions with laser beam diameter on the target surface. The minimum size of the laser beam's diameter depends on the focal length, which changes the laser's fluence. As a result, nanoparticles are created with multiple shapes and sizes. At the same time, if the diameter of the laser beam is small, the size of the nanoparticles becomes uniform, and its diameter is about 2 to 5 nm. If the diameter of the laser beam is too large, the size of the nanoparticles is enlarged, and the average size is about 20 nm [42].

Nath et al. [64] changed the position of the target lens under different focal conditions (up, down, and at the focal point) in **Figure 24**. They showed that focal condition is an important parameter for synthesizing small-size nanoparticles with

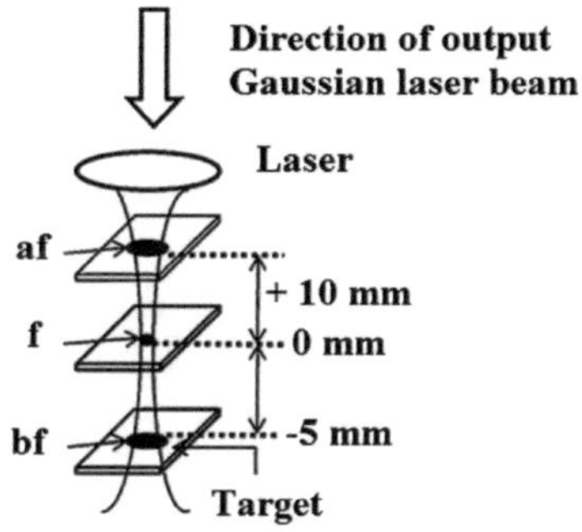

Figure 24.
Changing the diameter of the laser beam to above the focal point (af), the focal point (f), and below the focal point (bf) [64].

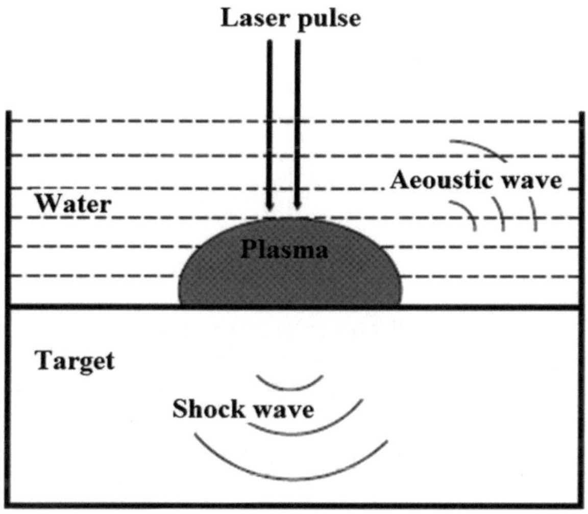

Figure 25.
The design shows the propagation of sound waves in the enclosed liquid (liquid and shock waves on the solid target) [65].

beam distribution. This factor changes the laser fluence and the degree of ionization of the liquid environment containing colloids.

A regular plasma plume can be detected during the laser radiation. Severe plasma will be created if the target is below the focal point. At low laser energy, the target temperature increases, but its evaporation rate decreases due to the liquid's function as a cooling agent on its surface, eliminating the formation of plumes. At medium laser beam energy, the plume is formed slowly. While at high energy, target laser ablation in the liquid environment results in the production of a plasma plume, which is visible to the naked eye near the target surface and creates a significant sound that can be due to the breaking bubble caused by cavitation from the evaporation of the fluid layer that is in close contact with the plasma. The release of the plasma and related sounds near the focal point are much greater [23].

According to **Figure 25**, a plasma plume can be seen during the radiation. When the laser is performed in a liquid environment, the fluid-induced enclosure on the solid target produces a shock wave in the plasma plume. As shown in **Figure 25**, evaporated species are classified as highly aroused ionic particles that are inappropriately abandoned in their primary quantum modes and emit electromagnetic radiation. It has been shown that the most severe plasma is produced by placing the target at a point a little before the geometric focal point [65].

5.7 Effect of scan speed

The interaction of laser pulses with plasma, bubbles, and laser-caused particles limits the efficiency of nanoparticles in the PLAL. It is only possible to reduce the protective effects caused by self-induced plasma by using shorter pulse times and changing the scan speed to minimize the effect caused by previous laser pulses. The spatial separation strategy of consecutive laser pulses is controlled by adjusting the interaction of the laser point size, the scan speed (the relative movement between the laser beam and the target), and the pulse repetition rate to prevent previous pulse protection [23].

Sattari et al. [66] found that the Al_2O_3 target PLAL efficiency can be increased by reducing the spatial overlap of consecutive laser pulses by increasing the scan speed in the IR laser with the ns pulse. The researchers reached the maximum efficiency by completely separating the pulses from each other. However, the linear decrease in efficiency was achieved at much higher distances due to the less target heat heating. Wagener et al. [67] studied the distance between the optical pulses in the laser ablation with the VIS laser with a 7 ps pulse of zinc metal target in tetrahydrofuran. However, reducing efficiency at distances between pulses was not more noticeable without overlapping. This was probably due to these researchers' lower pulse time (ps pulse). Streubel et al. [18] showed no efficiency decreases between longer pulse distances during 3 ps pulse time.

The thermal heating of the ablation target during the longer pulses will affect the distance between the optimal pulses because the heating mechanism is more prevalent in longer pulses. In short, the PLAL efficiency can be optimized by performing a complete pulse separation. For longer pulse time, the efficiency between the optimum pulses is maximized. **Figure 26** shows the effects of the distance between the appropriate pulses for the production of nanoparticles by the PLAL method.

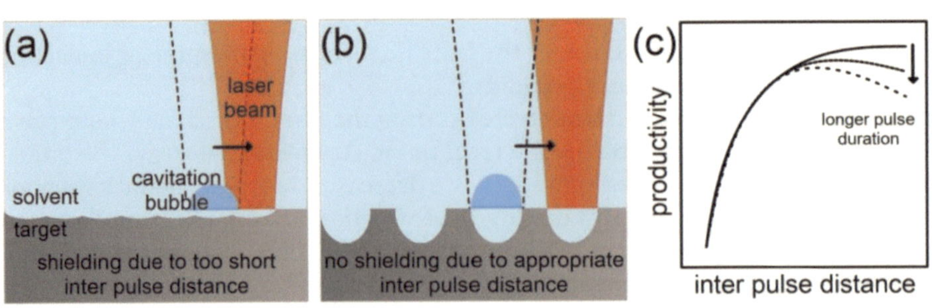

Figure 26.
(a) The distance image between the very short pulse, (b) the distance between the appropriate pulse in the metal laser ablation in liquids, and (c) the graph of the relative effects of the laser pulse time on the dependence of the efficiency on the distance between the pulses [23].

6. Effect of ablation environment on nanoparticles produced

6.1 Effect of liquid environment on the size of nanoparticles

The size of nanoparticles in the liquid after ablation in the liquid phase can be controlled with parameters such as laser fluence, pulse width and pulse repetition rate, wavelength, focal length, and ablation environment. In addition, the average size of colloidal nanoparticles is inversely related to fluid depth [57]. Increased liquid temperature changes the morphology of the nanoparticles from spherical to long [68].

Control of nanoparticles is an essential aspect of colloidal synthesis because the physical and chemical properties of the metal nanoparticles depend very much on their size. The physical and chemical properties of liquids and stabilizers in the liquid environment affect the synthesized nanoparticles. The viscosity, density, and fluid surface tension affect the bubble caused by cavitation and the enclosure of the plasma plume. Increasing viscosity in the liquid environment increases ablation efficiency by improving the plasma plume enclosure. Also, reducing the accumulation causes nanoparticle stability in the liquid [27].

Stabilizers change viscosity, density, and liquid surface tension, which affect the bubble dynamics caused by cavitation and enclosure of the plasma plume on the target. An example of this is the silver nanoparticle colloids prepared in deionized water (DW), ethanol, and polyvinylpyrrolidone (PVP) *via* laser ablation with a wavelength of 1064 nm to determine the effect of the liquid environment on the average size of nanoparticles. The nanoparticle distribution results and ablation efficiency results are presented in **Table 1**. Changing the fluid environment in the DW, ethanol, and PVP leads to changes in the average size of nanoparticles [57].

Also, the effect of changing the physical-chemical properties of liquid is associated with the interactions of nanoparticles and stabilizers. **Table 2**, for example, shows the physical properties of DW and acetone as two commonly used liquid environments in the PLAL process [69, 70]. **Figure 27** shows the interaction between stabilizers and ablated materials in different degrees. In studies on the effect of SDS, researchers concluded that SDS anions have electrostatic interactions with nanoparticles and form a molecular layer that restricts the accumulation and growth of nanoparticles by absorbing free atoms [71]. For example, the size of the gold nanoparticles from 20 nm in pure water is reduced to about 10 nm in a 10 mM SDS aqueous solution [59]. As shown in **Figure 27(a)**, reducing the size of nanoparticles during growth is only effective if the concentration of ligands is large enough to overlap the nanoparticles and prevent the accumulation and deposition of nanoparticles. Similar effects are also

Liquid environment	The average size of nanoparticles	Ablation efficiency	Distribution of nanoparticle size
PVP	16	High	Narrow
DW	23	Low	Wide
Ethanol	26	Very low	Narrow

Table 1.
Average size, ablation efficiency, and colloidal silver nanoparticle size distribution in three different PVP, DW, and ethanol environments [57].

Parameters	Acetone	DW
Dielectric constant	21.01	80.10
Boiling point (°C)	56	100
Density ($g.ml^{-1}$)	0.7845	0.9970
Special heat capacity on 25°C ($J.g.K^{-1}$)	2.175	4.180
Surface tension in 25°C ($mN.m^{-1}$)	22.72	71.99
Thermal conductive on 25°C ($Wm.K^{-1}$)	0.161	0.6062
Dipole moment (D)	2.88	1.8546
Viscosity on 25°C (mPa.S)	0.306	0.890
Reduction potential (V)	−1.2	−0.83
The refractive index in wavelength of 1064 nm	1.36135	1.32604
The passage of light (thickness of 1 cm)	≈1	0.54559

Table 2.
Physical and chemical properties of distilled water and acetone [69, 70].

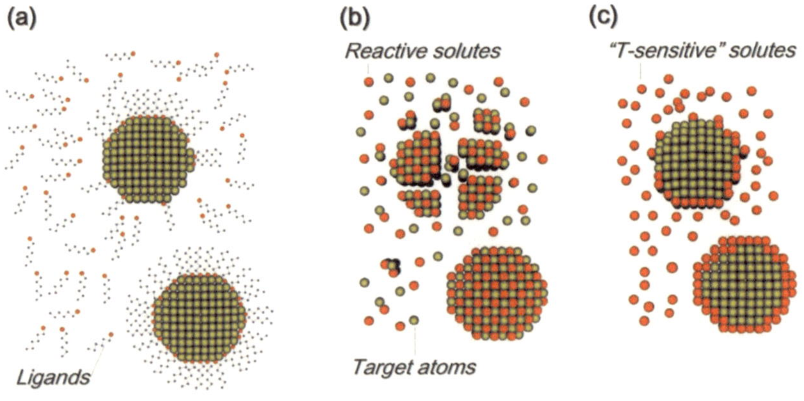

Figure 27.
Effect of stabilizers on nanoparticles (a) preventing growth, accumulation, and deposition of nanoparticles with physical or chemical interactions of ligands on the nanoparticle surface, (b) combining stabilizers with ablated materials and formation of new composition, and (c) the nucleation place of nanoparticles on stabilizers [27].

observed with stabilizers such as cyclodextrins, biopolymers, PVP, or cetyltrimethylammonium bromide (CTAB) [27]. As shown in **Figure 27(b)**, some molecules change the composition of nanoparticles at the highest level of interaction between stabilizers and ablated materials. For example, gold laser ablation in NaCl aqueous solution causes Au-Cl chemical bonds. In the case of NaCl, creating a high zeta potential increases surface repulsion among nanoparticles and creates smaller particles than pure water [45]. Similar results have been reported for silver laser ablation with NIR pulses [72]. In addition, as shown in **Figure 27(c)**, if spontaneous or temperature chemical reactions activate the solution, they can act as nanoparticle nucleation sites. For example, noble metal salts on the metal nanoparticles are reduced during the laser ablation process to create core-shell structures [27, 73].

Applying the magnetic field during the laser ablation process also increases the concentration and size of the nanoparticles, thereby increasing the absorption and efficiency. Using the external electric field decreases the size of the resulting nanoparticles, and tin and gold samples created under different electric fields can be mentioned [74].

6.2 Effect of the fluid environment on the morphology of the nanoparticles

It has also been found that liquids such as ethanol, DW, and acetone affect the morphology of the synthesized nanoparticles in the laser ablation method. The liquid controls the morphology, size, and distribution of nanoparticles and their composition. Its effect can be seen in **Figure 28** on the average size, distribution, and morphology of the tin produced by the PLAL [75]. The dipole moment in liquid is essential for producing smaller nanoparticles. The dipole moment is much higher for the acetone environment than DW and ethanol. For this reason, acetone's significant dipole moment leads to the smaller size of the tin nanoparticles because it prevents the cluster from growing in the plume. Therefore, the likelihood of spherical nanoparticles also increases [75].

Laser ablation for the production of nanoparticles relies on two mechanisms: (1) direct nucleation of atoms in the dense plume and (2) the act of nanoparticles as a growth center for the production of new nanoparticles. The effect of these mechanisms results in the distribution of the wide size of the nanoparticles. Compact and

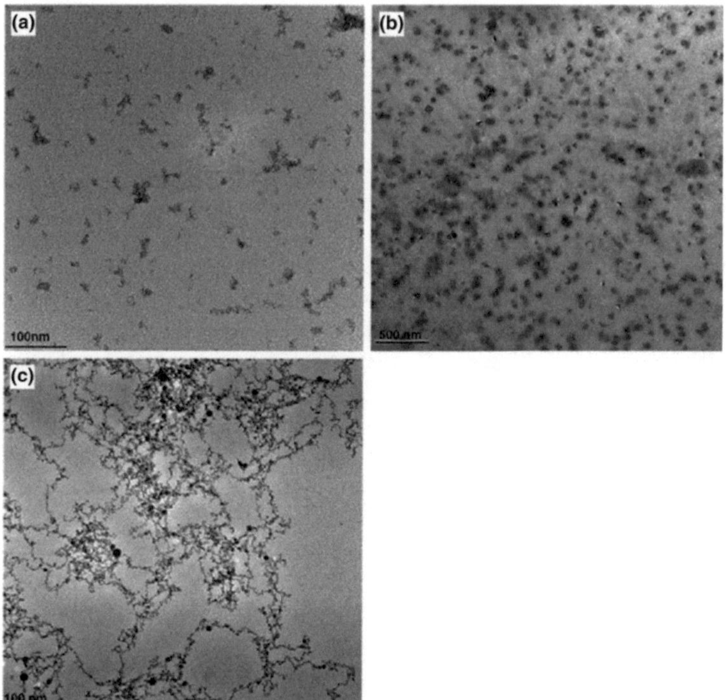

Figure 28.
TEM images of nanoparticles synthesized in the laser fluence 2.3 J.cm^{-2} in (a) ethanol, (b) DW, and (c) acetone [75].

Liquid	Dipole moment (D)	Average size of particles (nm)	Size distribution (nm)	Morphology
Ethanol	1.69	—	—	Filamentous state
DW	1.85	37	±10	Stretched state
Acetone	2.89	2	±1	Spherical state

Table 3.
Nanoparticle size, size distribution, and morphology were observed for tin nanoparticles prepared by laser ablation in different liquid environments [75].

strong bonds are absorbed onto the surface by very polar molecules. The electrostatic repulsion force of the dual electrical layers on the overlaps of the nuclei and the clusters in the plume prevents more growth, accumulation, or deposition. The average size, size distribution, and shape of particles obtained in different liquid environments are presented in **Table 3**.

Azawi et al. [57] synthesized metal nanoparticles with spherical morphology by controlling the process parameters in water, ethanol, and PVP aqueous solutions at a concentration of 20 mM. Aye et al. [76] also concluded that synthesized nanoparticles' morphology and physical and chemical properties can be regulated by controlling laser ablation parameters in the liquid environment. For example, when using copper as a target, immersing it in acetone instead of water at low laser fluence results in smaller nanoparticles and non-oxidation [23]. In addition, it is said that laser ablation in a stirring liquid environment is more effective than in a static liquid environment and increases ablation efficiency (up to 30%). The quality of colloidal properties is achieved [77].

7. The composition of the target matter in the process of ablation

All quantities that affect laser ablation can be somehow related. The characteristics of the target exposed to radiation significantly influence the ablation process. This material has a distinct melting temperature, and the melting process is initiated only when the laser energy is capable of raising the surface temperature of the material to this threshold. For example, the melting temperature of the copper is about 1358 K. If the laser with a power of 102 $W.cm^{-2}$ is used, the surface temperature can reach 1500 K, which is not too much to produce a significant amount of molten layer on the surface. However, a laser with a power of 108 $W.cm^{-2}$ reaches a surface temperature of 8000 K and converts about 1.8 μm of the target surface to a liquid phase. It is undoubtedly different in other cases. Other quantities, such as boiling temperature, special heat capacity, and heat conductivity of the target matter, are important in the ablation process. On the other hand, nanoparticles produced with the composition of the target matter have the maximum absorption at a specific wavelength, and the composition of nanoparticles can be found even by examining the nanoparticle absorption spectrum [42]. For example, **Table 4** shows the physical properties of copper and silver [69, 70, 78–80]. In this table, the values of the refractive index (n) and the extinction coefficient (k) indicate that the more k> > n in metals show more shine (higher reflection). On the other hand, if the values of $k \approx n \approx 3$ are divided into

Physical parameters	Silver	Copper	Reference
Surface energy ($J.m^{-2}$)	1.25	1.825	[78]
Electrons-phonon coupling constant (λ)	0.1	0.15	[79]
Refractive index (500 nm)	0.13	1.13	[80]
Refractive index (1000 nm)	0.21	0.33	[80]
Extinction coefficient (500 nm)	2.92	2.56	[80]
Extinction coefficient (1000 nm)	6.76	6.6	[80]
Melting point (°C)	961.8	1.85	[69]
Boiling point (°C)	2162	2562	[69]
Thermal conductivity ($Wcm.K^{-1}$)	4.29	4.01	[69]
Reduction potential (V)	(Ag^{+}) 0.8 and (Ag^{2+}) 1.98	(Cu^{+}) 0.52 and (Cu^{2+}) 0.34	[70]

Table 4.
Physical and chemical properties of copper and silver.

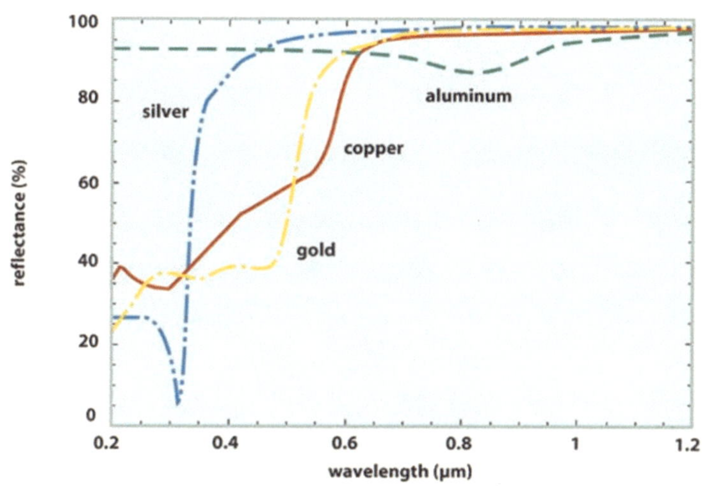

Figure 29.
Reflection of some shiny metal in the in the wavelengths of 200 to 1200 nm [80].

gray metals. For example for silver in the wavelength of 500 nm is n = 0.13 and K = 2.92, known as shiny metal. However, tungsten in this wavelength (500 nm) is n = 3.4 and k = 2.69, known as gray metal. **Figure 29** shows the reflection of some shiny metals (copper, silver, gold, and aluminum) in the wavelengths of 200 to 1200 nm [80].

8. Conclusion

In this study, different parameters of the PLAL process for the preparation of metal nanoparticles were examined. Higher laser fluence results in an increase in nanoparticle production efficiency. However, further increased laser fluence reduces metal target ablation due to its effect of self-absorption of nanoparticles. Laser pulse time, by increasing the pulse width from the ps and fs laser to the ns laser, increases efficiency due to the larger cavity in the target and the size of the smaller nanoparticles due to the maximum evaporation of the nanoparticles in the plasma plume. As the repetition rate (frequency) increases of laser pulses, due to the effect of higher heat density, the efficiency of the target increases, and the size of the nanoparticles is reduced due to the crushing of nanoparticles due to successive pulses. Also, with increased wavelengths from UV to NIR, the efficiency and size of nanoparticles will increase due to the self-absorption of nanoparticles. It should be noted that due to the interactions of the laser parameters and the liquid environment of the synthesis, it is difficult to have a conclusion about the efficiency and size of the nanoparticles, and this is a general result according to the results of most researchers, which are mentioned in the sources. An essential key to the synthesis of nanoparticles by the PLAL method is to optimize the process parameters.

Conflict of interest

The authors declare no conflict of interest.

Author details

Hamed Naderi-Samani and Reza Shoja Razavi*
Faculty of Materials and Manufacturing Technologies, Malek Ashtar University of Technology, Tehran, Iran

*Address all correspondence to: shoja_r@yahoo.com

References

[1] Zhang D, Gokce B, Barcikowski S. Laser synthesis and processing of colloids: Fundamentals and applications. Chemical Reviews. 2017;**117**(5): 3990-4103

[2] Amendola V et al. Room-temperature laser synthesis in liquid of oxide, metal-oxide core-shells, and doped oxide nanoparticles. Chemistry–A European Journal. 2020;**26**(42):9206-9242

[3] Amans D, Cai W, Barcikowski S. Status and Demand of Research to Bring Laser Generation of Nanoparticles in Liquids to Maturity. Vol. 488. Netherlands: Elsevier; 2019. pp. 445-454

[4] Ziefuß AR et al. Laser fragmentation of colloidal gold nanoparticles with high-intensity nanosecond pulses is driven by a single-step fragmentation mechanism with a defined educt particle-size threshold. The Journal of Physical Chemistry C. 2018;**122**(38): 22125-22136

[5] Meader VK, John MG, Frias Batista LM, Ahsan S, Tibbetts KM. Radical chemistry in a femtosecond laser plasma: Photochemical reduction of Ag+ in liquid ammonia solution. Molecules. 2018;**23**(3):532

[6] Hunter BM, Blakemore JD, Deimund M, Gray HB, Winkler JR, Muller AM. Highly active mixed-metal nanosheet water oxidation catalysts made by pulsed-laser ablation in liquids. Journal of the American Chemical Society. 2014;**136**(38):13118-13121

[7] Marzun G, Levish A, Mackert V, Kallio T, Barcikowski S, Wagener P. Laser synthesis, structure and chemical properties of colloidal nickel-molybdenum nanoparticles for the substitution of noble metals in heterogeneous catalysis. Journal of Colloid and Interface Science. 2017;**489**: 57-67

[8] Ma C, Yan J, Huang Y, Yang G. Directional scattering in a germanium nanosphere in the visible light region. Advanced Optical Materials. 2017;**5**(24): 1700761

[9] Niyuki R et al. Double threshold behavior in a resonance-controlled ZnO random laser. APL Photonics. 2017;**2**(3): 036101

[10] Petersen S, Barcikowski S. In situ bioconjugation: Single step approach to tailored nanoparticle-bioconjugates by ultrashort pulsed laser ablation. Advanced Functional Materials. 2009; **19**(8):1167-1172

[11] Hess C et al. Dose-dependent surface endothelialization and biocompatibility of polyurethane noble metal nanocomposites. Journal of Biomedical Materials Research Part A. 2014;**102**(6): 1909-1920

[12] Zhang D, Gökce B. Perspective of laser-prototyping nanoparticle-polymer composites. Applied Surface Science. 2017;**392**:991-1003

[13] Maurer E, Barcikowski S, Gökce B. Process chain for the fabrication of nanoparticle polymer composites by laser ablation synthesis. Chemical Engineering & Technology. 2017;**40**(9): 1535-1543

[14] Compagnini G, Scalisi AA, Puglisi O. Ablation of noble metals in liquids: A method to obtain nanoparticles in a thin polymeric film. Physical Chemistry Chemical Physics. 2002; **4**(12):2787-2791

[15] Jonušauskas L et al. Plasmon assisted 3D microstructuring of gold nanoparticle-doped polymers. Nanotechnology. 2016;**27**(15):154001

[16] Doñate-Buendía C et al. Oxide dispersion-strengthened alloys generated by laser metal deposition of laser-generated nanoparticle-metal powder composites. Materials & Design. 2018; **154**:360-369

[17] Hupfeld T et al. A new approach to coat PA12 powders with laser-generated nanoparticles for selective laser sintering. Procedia CIRP. 2018;**74**: 244-248

[18] Streubel R, Barcikowski S, Gökce B. Continuous multigram nanoparticle synthesis by high-power, high-repetition-rate ultrafast laser ablation in liquids. Optics Letters. 2016;**41**(7): 1486-1489

[19] Jendrzej S, Gökce B, Epple M, Barcikowski S. How size determines the value of gold: Economic aspects of wet chemical and laser-based metal colloid synthesis. ChemPhysChem. 2017;**18**(9): 1012-1019

[20] Tymoczko A et al. How the crystal structure and phase segregation of Au–Fe alloy nanoparticles are ruled by the molar fraction and size. Nanoscale. 2018; **10**(35):16434-16437

[21] Wagener P et al. Solvent-surface interactions control the phase structure in laser-generated iron-gold core-shell nanoparticles. Scientific Reports. 2016; **6**(1):1-12

[22] Du H, Castaing V, Guo D, Viana B. Rare-earths doped-nanoparticles prepared by pulsed laser ablation in liquids. Ceramics International. 2020; **46**(16):26299-26308

[23] Fazio E et al. Nanoparticles engineering by pulsed laser ablation in liquids: Concepts and applications. Nanomaterials. 2020;**10**(11):2317

[24] Blatchford C-G, Campbell J, Creighton JA. Plasma resonance—Enhanced Raman scattering by absorbates on gold colloids: The effects of aggregation. Surface Science. 1982; **120**(2):435-455

[25] Nano, Nanoparticle Production by Pulse Laser Ablation in Liquid. IMRA. Available from: http://nano.imra.com/ [Accessed: June 13, 2022]

[26] Particular, Nanoparticle Production by Pulse Laser Ablation in Liquid. Particular GmbH. Available from: http://particular.eu/start.html [Accessed: June 13, 2022]

[27] Amendola V, Meneghetti M. What controls the composition and the structure of nanomaterials generated by laser ablation in liquid solution? Physical Chemistry Chemical Physics. 2013;**15**(9): 3027-3046

[28] Mat Isa SZ, Zainon R, Tamal M. State of the art in gold nanoparticle Synthesisation via pulsed laser ablation in liquid and its characterisation for molecular imaging: A review. Materials. 2022;**15**(3):875

[29] De Giacomo A et al. Cavitation dynamics of laser ablation of bulk and wire-shaped metals in water during nanoparticles production. Physical Chemistry Chemical Physics. 2013;**15**(9): 3083-3092

[30] Dell'Aglio M, Motto-Ros V, Pelascini F, Gornushkin IB, De Giacomo A. Investigation on the material in the plasma phase by high temporally and spectrally resolved emission imaging during pulsed laser ablation in liquid

(PLAL) for NPs production and consequent considerations on NPs formation. Plasma Sources Science and Technology. 2019;**28**(8): 085017

[31] Taccogna F, Dell'Aglio M, Rutigliano M, Valenza G, De Giacomo A. On the growth mechanism of nanoparticles in plasma during pulsed laser ablation in liquids. Plasma Sources Science and Technology. 2017;**26**(4): 045002

[32] Dell'Aglio M, De Giacomo A. Plasma charging effect on the nanoparticles releasing from the cavitation bubble to the solution during nanosecond pulsed laser ablation in liquid. Applied Surface Science. 2020;**515**:146031

[33] Kanitz A, Kalus M, Gurevich E, Ostendorf A, Barcikowski S, Amans D. Review on experimental and theoretical investigations of the early stage, femtoseconds to microseconds processes during laser ablation in liquid-phase for the synthesis of colloidal nanoparticles. Plasma Sources Science and Technology. 2019;**28**(10):103001

[34] Tanabe R, Nguyen TT, Sugiura T, Ito Y. Bubble dynamics in metal nanoparticle formation by laser ablation in liquid studied through high-speed laser stroboscopic videography. Applied Surface Science. 2015;**351**: 327-331

[35] Shih C-Y et al. Two mechanisms of nanoparticle generation in picosecond laser ablation in liquids: The origin of the bimodal size distribution. Nanoscale. 2018;**10**(15):6900-6910

[36] Reich S et al. Fluence threshold behaviour on ablation and bubble formation in pulsed laser ablation in liquids. ChemPhysChem. 2017;**18**(9): 1084-1090

[37] Strutt JW. LVIII. On the scattering of light by small particles. The London, Edinburgh, and Dublin Philosophical Magazine and Journal of Science. 1871; **41**(275):447-454

[38] Creighton JA, Eadon DG. Ultraviolet–visible absorption spectra of the colloidal metallic elements. Journal of the Chemical Society, Faraday Transactions. 1991;**87**(24):3881-3891

[39] Chewchinda P, Tsuge T, Funakubo H, Odawara O, Wada H. Laser wavelength effect on size and morphology of silicon nanoparticles prepared by laser ablation in liquid. Japanese Journal of Applied Physics. 2013;**52**(2R):025001

[40] Nedialkov NN, Atanasov P, Sawczak M, Sliwinski G. Ablation of ceramics with ultraviolet, visible, and infrared nanosecond laser pulses. In: XIV International Symposium on Gas Flow, Chemical Lasers, and High-Power Lasers. Vol. 5120. USA: SPIE; 2003. pp. 703-708

[41] Sikora A, Grojo D, Sentis M. Wavelength scaling of silicon laser ablation in picosecond regime. Journal of Applied Physics. 2017;**122**(4):045702

[42] Bogaerts A, Chen Zh. Effect of laser parameters on laser ablation and laser-induced plasma formation: A numerical modeling investigation. Spectrochimica Acta Part B: Atomic Spectroscopy. 2005; **60**(9-10):1280-1307

[43] Solati E, Mashayekh M, Dorranian D. Effects of laser pulse wavelength and laser fluence on the characteristics of silver nanoparticle generated by laser ablation. Applied Physics A. 2013;**112**(3):689-694

[44] Tsuji T, Okazaki Y, Tsuji M. Photo-induced morphological conversions of

silver nanoparticles prepared using laser ablation in water—Enhanced morphological conversions using halogen etching. Journal of Photochemistry and Photobiology A: Chemistry. 2008;**194**(2–3, 253):247

[45] Amendola V, Meneghetti M. Laser ablation synthesis in solution and size manipulation of noble metal nanoparticles. Physical Chemistry Chemical Physics. 2009;**11**(20): 3805-3821

[46] Patra N et al. Parametric investigations on the influence of nano-second Nd3+: YAG laser wavelength and fluence in synthesizing NiTi nano-particles using liquid assisted laser ablation technique. Applied Surface Science. 2016;**366**:104-111

[47] Chaturvedi A, Joshi M, Mondal P, Sinha A, Srivastava A. Growth of anatase and rutile phase TiO2 nanoparticles using pulsed laser ablation in liquid: Influence of surfactant addition and ablation time variation. Applied Surface Science. 2017;**396**:303-309

[48] Desarkar HS, Kumbhakar P, Mitra A. Optical properties of tin oxide nanoparticles prepared by laser ablation in water: Influence of laser ablation time duration and laser fluence. Materials Characterization. 2012;**73**:158-165

[49] Badr Y, Mahmoud M. Excimer laser photofragmentation of metallic nanoparticles. Physics Letters A. 2007; **370**(2):158-161

[50] Giorgetti E et al. TiO2 nanoparticles obtained by laser ablation in water: Influence of pulse energy and duration on the crystalline phase. Journal of Alloys and Compounds. 2015;**643**:S75-S79

[51] Schwenke A, Wagener P, Nolte S, Barcikowski S. Influence of processing time on nanoparticle generation during picosecond-pulsed fundamental and second harmonic laser ablation of metals in tetrahydrofuran. Applied Physics A. 2011;**104**(1):77-82

[52] Momma C et al. Short-pulse laser ablation of solid targets. Optics Communications. 1996;**129**(1-2, 142):134

[53] Sylvestre J-P, Poulin S, Kabashin AV, Sacher E, Meunier M, Luong JH. Surface chemistry of gold nanoparticles produced by laser ablation in aqueous media. The Journal of Physical Chemistry B. 2004;**108**(43):16864-16869

[54] Sylvestre J-P, Kabashin AV, Sacher E, Meunier M. Femtosecond laser ablation of gold in water: Influence of the laser-produced plasma on the nanoparticle size distribution. Applied Physics A. 2005;**80**(4):753-758

[55] Valverde-Alva M et al. Laser ablation efficiency during the production of Ag nanoparticles in ethanol at a low pulse repetition rate (1–10 Hz). Laser Physics Letters. 2016; **13**(10):106002

[56] Tsuji T, Tsuboi Y, Kitamura N, Tsuji M. Microsecond-resolved imaging of laser ablation at solid–liquid interface: Investigation of formation process of nano-size metal colloids. Applied Surface Science. 2004;**229**(1-4):365-371

[57] Al-Azawi MA et al. The effects of the ambient liquid medium on the ablation efficiency, size and stability of silver nanoparticles prepared by pulse laser ablation in liquid technique. Jurnal Teknologi. 2016;**78**(3):7-11

[58] Piriyawong V, Thongpool V, Asanithi P, Limsuwan P. Effect of laser pulse energy on the formation of alumina nanoparticles synthesized by

laser ablation in water. Procedia Engineering. 2012;**32**:1107-1112

[59] Mafuné F, Kohno J-y, Takeda Y, Kondow T, Sawabe H. Formation of gold nanoparticles by laser ablation in aqueous solution of surfactant. The Journal of Physical Chemistry B. 2001; **105**(22):5114-5120

[60] Kabashin AV, Meunier M. Synthesis of colloidal nanoparticles during femtosecond laser ablation of gold in water. Journal of Applied Physics. 2003; **94**(12):7941-7943

[61] Nichols WT, Sasaki T, Koshizaki N. Laser ablation of a platinum target in water. II. Ablation rate and nanoparticle size distributions. Journal of Applied Physics. 2006;**100**(11):114911

[62] Chewchinda P et al. Preparation of Si nanoparticles by laser ablation in liquid and their application as photovoltaic material in quantum dot sensitized solar cell. Journal of Physics: Conference Series. 2014;**518**(1):012023

[63] Elsayed KA, Imam H, Ahmed M, Ramadan R. Effect of focusing conditions and laser parameters on the fabrication of gold nanoparticles via laser ablation in liquid. Optics & Laser Technology. 2013;**45**:495-502

[64] Nath A, Laha S, Khare A. Effect of focusing conditions on synthesis of titanium oxide nanoparticles via laser ablation in titanium–water interface. Applied Surface Science. 2011;**257**(7): 3118-3122

[65] Messina E. Metal Nanoparticles Produced by Pulsed Laser Ablation in Liquid Environment. Italy: Universita'degli Studi di Catania; 2011

[66] Sattari R, Sajti C, Khan S, Barcikowski S. Scale-up of nanoparticle production during laser ablation of ceramics in liquid media. In: International Congress on Applications of Lasers & Electro-Optics. Vol. 2008. USA: Laser Institute of America; 2008. p. N204

[67] Wagener P, Schwenke A, Chichkov BN, Barcikowski S. Pulsed laser ablation of zinc in tetrahydrofuran: Bypassing the cavitation bubble. The Journal of Physical Chemistry C. 2010; **114**(17):7618-7625

[68] Al-Dahash G, Obaid NM, Majeed HA. The effect of liquid environment and magnetic field on optical properties of Pt nanoparticles colloidal prepared by pulsed laser ablation. International Journal of ChemTech Research. 2016;**9**:118-130

[69] Haynes WM. CRC Handbook of Chemistry and Physics. USA: CRC Press; 2016

[70] Charlot G. Selected Constants Oxydo-Reduction Potentials: Tables of Constants and Numerical Data Affiliated to the International Union of Pure and Applied Chemistry. Vol. 8. Turkiye: Elsevier; 2013

[71] Baiee RM. Generation of Ultra-Fine Nanoparticles by Laser Ablation in Liquid. United Kingdom: The University of Manchester; 2019

[72] Šišková K, Vlckova B, Turpin P, Fayet C. Ion-specific effects on laser ablation of silver in aqueous electrolyte solutions. The Journal of Physical Chemistry C. 2008;**112**(12):4435-4443

[73] Jimenez E, Abderrafi K, Abargues R, Valdes JL, Martinez-Pastor JP. Laser-ablation-induced synthesis of SiO2-capped noble metal nanoparticles in a single step. Langmuir. 2010;**26**(10): 7458-7463

[74] Sapkota D, Li Y, Musaev OR, Wrobel JM, Kruger MB. Effect of electric fields on tin nanoparticles prepared by laser ablation in water. Journal of Laser Applications. 2017;**29**(1):012002

[75] Bajaj G, Soni R. Effect of liquid medium on size and shape of nanoparticles prepared by pulsed laser ablation of tin. Applied Physics A. 2009; **97**(2):481-487

[76] Aye HL, Choopun S, Chairuangsri T. Influence of solvents on characteristics of nanoparticles prepared by pulsed laser ablation on iron target. Chiang Mai University Journal of Natural Sciences. 2014;**31**:37-42

[77] Resano-Garcia A, Battie Y, Koch A, En Naciri A, Chaoui N. Influence of the laser light absorption by the colloid on the properties of silver nanoparticles produced by laser ablation in stirred and stationary liquid. Journal of Applied Physics. 2015;**117**(11):113103

[78] Vitos L, Ruban A, Skriver HL, Kollár J. The surface energy of metals. Surface Science. 1998;**411**(1–2):186-202

[79] Grimvall G. The electron-phonon interaction in normal metals. Physica Scripta. 1976;**14**(1–2):63

[80] Photonics marketplace, refractive index, mirrors: Coating choice makes a difference. Available from: https://www.photonics.com/Articles/Mirrors_Coating_Choice_Makes_a_Difference/a25501 [Accessed: June 14, 2020]

Chapter 5

Ultrashort-Pulse Burst-Mode Materials Processing and Laser Surgery

Robin S. Marjoribanks, Jinseng Tang, Thomas Dzelzainis, Melissa Prickaerts, Lothar Lilge, Margarete Akens, Colin Veevers, Nick N. Gharabaghi, Andreas Hitzler, Seydi Yavas and Sohret Görkem Karamuk

Abstract

Laser processing of materials and biological tissues has evolved in stages, ever since the earliest use of the laser for gross deposition of heat and for ablation. For instance, wavelength specificity was an early development that facilitated the treatment of certain biological tissues, while leaving others relatively unaffected. Ultrashort-pulse material ablation escapes the usual paradigm of heat diffusion because of the comparisons of scales: A rarefaction wave can cut through the thin layer of femtosecond-laser-heated material and carry away the absorbed energy before much heat can diffuse into the substrate. Burst-mode femtosecond laser ablation brings yet another paradigm, in which the laser fluence is divided over two disparate timescales: the ultrashort duration of a pulse and the microsecond-scale duration of a burst. This division of timescales opens new avenues for control, because much of the governing physics is about the comparison of timescales—for instance, the timescale of thermalization of heated electrons into the substrate lattice or the timescale of hydrodynamic ablation. Applications to fused silica, to *in vitro* cell-cultures prepared in hydrogels, and to *ex vivo* articular cartilage help to show what is different in the science of ultrashort-pulse burst-mode laser processing.

Keywords: ultrashort pulse, burst-mode laser, femtosecond laser, laser-materials processing, laser surgery, plasma-mediated ablation, persistence of plasma, ablative quenching

1. Introduction

At the time of its invention, Irnee D'Haenens in Theodore Maiman's laboratory was credited with describing their first laser to Maiman as "a solution looking for a problem" [1]. Over the years, endless applications have been found, both directly and, perhaps with even greater impact, as an enabling technology.

In laser materials-processing and laser-surgery, this path toward problems solved might be usefully framed in terms of qualitative generations of applications that have re-framed what the laser can do:

First Generation—Power Delivery: cutting without tool bits, offering non-contact, low-contamination; spatially localized delivery of gross heat to workpieces, to cut or weld metals, cut cloth, wood and plastic. From lab to factory before the 1960s ended.

Second Generation—Wavelength-tunable selective absorption: differential control due to intrinsic or extrinsic absorbers; specialized cutting in plastics; surgical cauterization; port-wine-stain birthmark-removal. Roughly from the 1970s onward.

Third Generation—Modes of Fluence Delivery: controlling duration/intensity/pattern of pulses; ultrashort-pulse chirped-pulse amplification (CPA) lasers, low collateral damage surgery and micromachining, custom machining of microfluidics (emerging in the 1990s) and direct-write of internal optical waveguide structures in bulk glasses (emerging in the early 2000s).

At the most basic level, the final outcomes of laser treatment of materials or biological tissues are governed by several key observations:

1. At its simplest, material modification requires deposition of specific energy (energy/unit mass), a per-atom energy characteristic of chemical or material-structure changes, e.g., latent heat of melting, or of vaporization.

2. This deposited specific energy derives from a given absorbed laser fluence (*radiant exposure*: energy/unit area), together with a characteristic range of heat diffusion into the sample.

3. Therefore the most basic issues around laser materials-processing and laser-ablative surgery are the absorption and subsequent disposition of incident laser energy as heat.

In broad terms, the impact of long- and short-pulse laser materials-processing differ because of characteristic time-scales of materials and of processes, meaning that the two classes bring with them very different relationships to jobs needing to be done. Given that specific energy, and therefore laser fluence, is a driving figure-of-merit, we can use it to compare in generalities:

Long-pulse lasers—For a given delivered fluence, longer-duration pulses mean lower peak intensities. At lower intensities, nanosecond and microsecond-pulse lasers typically exploit linear absorption, either from intrinsic absorption and endogenous chromophores or from Joule heating in the case of conductors. Energy is absorbed over a depth determined by the inverse of the optical absorption coefficient. Dielectrics such as glass depend on nonlinear absorption and therefore high peak intensities in order to surpass the dielectric breakdown threshold; however, after breakdown and plasma creation the absorption becomes linear but largely independent of the original material. The result is that, for nanosecond or longer pulses on dielectrics, this runaway process necessarily means that far more energy than desired may be delivered.

Ultrashort-pulse lasers—For a given delivered fluence, peak intensity goes up as the pulse-duration shortens. Short-enough pulses then assure dielectric breakdown and plasma formation, regardless of whether a material has appreciable linear absorption. Absorption is then plasma-mediated, with the effect of making absorption "universal": independent of local variations in extrinsic absorption from defects, impurities or dopants, or independent of having distinct endogenous chromophores in different locations, as is the case for a differentiated biological tissue containing a mix of connective tissues, nerve and muscle.

In particular, plasma-mediated absorption, because of the effect of plasma screening, also puts a constraint on the optical absorption depth. This small optical absorption depth, and the limited amount of plasma produced by pulses that are over so quickly, mean that the mediating plasma that absorbs laser energy is thin. These are principal factors in why ultrashort laser-pulse processing leaves so little heat behind in the target.

Ultrashort-pulse lasers, used at intensities near the damage threshold, show much less shot-to-shot variability in outcome than do nanosecond and longer pulses, even in a pure dielectric such as fused silica [2, 3]. This, too, is due to a qualitative change in physics relationships governing ionization [4, 5]. For ultrashort-pulse lasers, shock-waves launched in the material also show a reduced range of shock-damage, a thing particularly important in laser-surgery [6, 7].

Ultrashort-pulse lasers are used to deliver fluence on target at values similar to those used for longer-pulse lasers, but they deliver that fluence over shorter timescales and at higher intensities. Each of these qualitative changes above traces back to a change in physics, obviously, but more particularly these changes trace to ways in which delivery by ultrashort-pulse lasers crosses characteristic timescales and intensities, and thereby qualitatively changes the governing physics-relationships.

2. Physics relationships of single-pulse ultrashort-pulse material processing

In physics, meaning naturally derives from relationship. Every formula relates parameters to an outcome, every plot shows a dependency. Dimensionless numbers mark for us, as ratios, the truly qualitative shifts in relationship that govern major changes in physical interactions. For instance, the Mach number marks the qualitative change that happens between subsonic and supersonic flight—the transition at which the control surfaces at the trailing edge of a wing can no longer affect airflow over the wing, because the pressure change induced by changes in, say, an aileron, cannot travel forward as a sound wave, against the oncoming flow of air which is traveling faster than the speed of sound. The Prandtl number Pr is the ratio between the diffusivity of momentum in a fluid flow to the diffusivity of heat in that fluid—it flags the flow of energy via bulk convection ($Pr > 1$) as compared to the flow principally via heat diffusion through a fluid.

For laser interactions with materials, **Figure 1** helps to give context to our relevant timescales. For instance, the extended plasma plume generated from laser ablation may take hundreds of nanoseconds to a microsecond to dissipate—it makes a few-nanosecond laser pulse look impulsive, like a Dirac delta-function, by comparison. Yet lateral thermal conduction or ambipolar diffusion in the plasma around critical density is so rapid that the rise and fall of a few-nanosecond pulse looks like an adiabatic change by comparison. However, ambipolar diffusion in the plasma, or lateral thermal conduction, seem frozen in time during the interaction of a subpicosecond pulse. Changing from few-nanosecond to subpicosecond laser pulses, the relationship of timescales between laser and material process is inverted—the role of those processes is qualitatively changed, and their significance is fundamentally different:

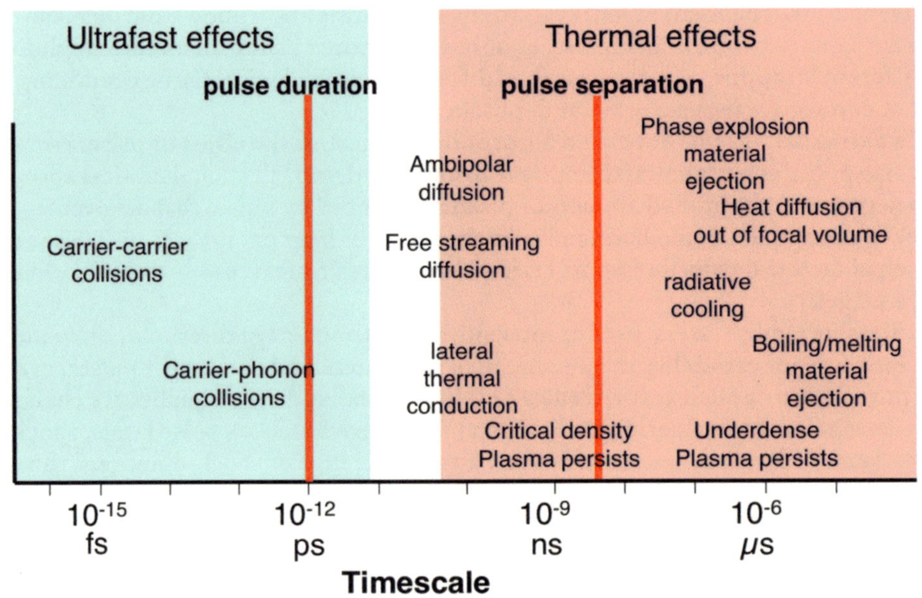

Figure 1.
Summary of material and process timescales relevant for laser materials-processing and surgery. For burst-mode ultrashort pulse laser processing, a new timescale is introduced by the time-separation of pulses within a burst. The repetition-rate period is also a timescale, typically long compared to these processes. The significance of the pulse duration or intraburst time delay is context-dependent, emerging from native timescales of the problem [8].

in the case of few-nanosecond pulses, ambipolar diffusion and lateral thermal conduction essentially follow the laser conditions in near steady-state; for subpicosecond laser pulses, they are virtually frozen in time.

In this general way, ultrashort pulse materials-processing is distinguished by a transition which changes laser-energy deposition from a near-steady state dynamic, with power deposited at a surface and then flowing onward by diffusion, to one in which a small thin volume is suddenly heated, and not long afterwards dispersed. Then whatever little heat has made it into the material, before the mediating plasma disperses, is left to diffuse inward after the fact, and its final impact may be determined long after the laser interaction has ended.

Among modes of fluence-delivery, femtosecond and few-picosecond laser pulses are unique in laser materials-processing and laser surgery for the small amount of collateral damage they typically produce in a target substrate, and the small amount of residual heat they leave behind [2, 3]. How does this categorically different result relate to the inversion of the relationship of timescales?

Fundamentally, free electrons cannot absorb energy from the passage of an infinite-plane oscillating electric field, but of course this changes with collisions. The principal path to plasma formation is avalanche ionization of atoms through collisions of electrons rapidly driven in the intense optical field. Yet collisional ionization by electrons requires that at least some free electrons be available, and original speculation was that in dielectrics like fused silica these were provided by multi-photon ionization [2, 3, 9]. Subsequent wavelength-scaling studies established that initial ionization originates instead from AC tunneling ionization of atoms in strong fields [4, 10]. For materials with intrinsic or extrinsic absorption, initial ionization can begin thermally, following initial linear absorption. This scenario is equally available to nanosecond interactions that create plasma. But there's a qualitative transition for collisional ionization under intense ultrashort-pulses.

The cross-section for atom ionization by laser-driven electron-impact depends importantly on the electric field strength: assuming constant fluence, the optical field strength going from few-nanosecond- to 100-femtosecond-duration pulses increases by orders of magnitude. Over this range, the cross-section for laser-driven collisions rises by orders and then trends toward saturation at higher field strengths [5, 11], fields routinely accessed by ultrashort-pulse lasers in laser materials-processing, but not typically by nanosecond or longer pulses. The consequence for laser materials-processing is that avalanche ionization is more sensitive to field-strength variations within the range of fields generated in nanosecond pulses than it is within the range of fields of subpicosecond pulses. Consequently, tests of damage threshold for ultrashort pulses are quite repeatable, whereas for longer pulses the range of scatter in laser intensity-values associated with damage is much greater [2, 3].

Once a thin mediating plasma has been established, absorption of laser energy into the plasma becomes quite generic and linear, and is largely determined by electron density and gradients of the electron density profile around the plasma critical-density surface.

2.1 Time scales for depth of heating in plasma-mediated absorption

Under these conditions, one of the relationships governing laser materials-processing is overthrown. During the time (i.e., the pulse duration τ) that laser energy is applied to a material or biological tissue, thermal diffusion produces a characteristic scale-length $L_{th} = \sqrt{D}$, where D is the thermal diffusion constant of the medium. This should not be understood as a range of the heat-affected zone, because it contains no reference to the absolute energy deposited, and, as noted above, a phase transition such as melting requires a specific energy, like the molar latent heat necessary for vaporization and ionization of a solid. Other heat-affected processes may be more complicated, and may depend on more than just the heat delivered or peak temperature achieved: an example would be protein-denaturation, which is governed by both temperature and duration, as expressed through the Arrhenius damage integral. In either case, the extent over which the threshold condition is met for being heat-affected is not the scale length of the distribution of heat. This thermal scale-length is instead a characteristic, much like the skin depth of an optical field penetrating a metal.

Indeed, the laser field has just such an optical skin-depth, δ, for the field (or an absorption scale-length for the intensity), in the thin plasma that has formed (*cf.* the Bouguer-Lambert law). The comparison of these two scale-lengths—thermal vs. absorption—is the next qualitative change in relationships that arises as the physics of interaction transitions, as progressively brief laser pulses effectively "freeze" important physical processes.

As pointed out by Pronko et al. [12], a wavelength of 800 nm produces an optical absorption scale-length of about 12 nm in metallic silver. The thermal diffusion scale-length L_{th} above will equal this in the case of a pulse of duration τ = 7 ps. On the timescale of pulses very much shorter than 7 ps, diffusion appears relatively "frozen," and the heat distribution during the laser pulse is set by the properties of the plasma alone. For substantially longer-duration pulses the heat scale-length during the pulse is governed by diffusion.

At the end of an ultrashort pulse, the tiny optical absorption depth of a dense plasma means a very thin layer of constant thickness has been heated to an ionized state. Therefore the material damage threshold is reached at a reduced laser fluence, which becomes independent of pulse duration.

For pulses tens of picoseconds and longer, a $\sqrt{\tau}$ dependence of damage-threshold fluence on irradiating-pulse duration was well-recognized [13–15]. For pulses shorter than about 10 ps, early studies from both Livermore and Michigan groups showed a relationship change, for damage in transparent dielectric targets [4, 5, 9]. Researchers initially disagreed about the behavior below this point of transition, but it's now largely agreed that the damage-threshold fluence does become constant for pulses with $\tau < 10$ ps.

More recently, refinements in modeling the non-thermal transport of electrons generated in intense laser-plasma interaction have added to the understanding of heat deposition in the substrate, including the range and heating of a fraction of electrons that penetrate directly [16, 17].

2.2 Time scales for erosion of thin mediating plasma layer

For femtosecond laser pulses, hydrodynamic expansion is a relatively slow process, and the mediating plasma layer is typically quite thin at the time the pulse ends. The absorbed energy resides as heat in this plasma layer, in close contact with the target substrate, and heat diffusion will carry a small amount of heat inward while the laser pulse is depositing energy into the absorption-depth layer.

A new time scale enters at this point, while the thin heated layer of plasma is still in thermal contact with the substrate. The subsequent erosion of this thin hot plasma layer is by a rarefaction wave—the "negative profile" hydrodynamic wave of loss of density, moving inward through the plasma layer. This rarefaction wave is the same general phenomenon as the process by which an inflated party balloon releases its internal overpressure once the latex rubber has been popped and very rapidly retracts over the whole surface, or by which a collapsed dam creates a traveling depletion wave in the water-level of the reservoir previously contained.

The timescale of erosion of the thin plasma layer is governed by the speed of propagation of a density wave in the plasma, c_s, together with the layer thickness d. After a time $t \cong d / c_s$, the heated plasma has decoupled from the target and been carried away in the expansion—the phenomenon of *ablative quenching*.

$$c_s = \sqrt{\frac{\bar{Z} k_B T_e}{M}} \qquad (1)$$

Here c_s is the ion-acoustic speed; T_e is the isothermal electron temperature. The ion temperature T_i is relatively low, so that $T_e \gg T_i$; $\bar{Z}$ is the ion average charge-state; M is the ion mass (the electrons are assumed to be isothermal, typically justified). An ion-acoustic wave is the plasma density wave in which the restoring force is derived from the electron pressure, via the electron temperature, and the inertial term is the ion mass. Effectively, the electrons dress the ion perturbations within a characteristic Debye length, coupling the electron and ion density perturbations together. See also Kerse et al. [18].

Thus, the time of heating is *not* the duration of the laser, but instead is the lifetime of the plasma layer in thermal contact with the target material. After this point, the heat diffusing inwardly in the target is only what energy had already propagated past the laser-heated plasma and into the substrate during the brief time the mediating plasma was in contact. For a subpicosecond pulse and an absorbed fluence of 40 J/cm^2, into a laser skin-depth of ~50 nm, a simple estimate of electron temperature gives an ablative-decoupling time of a few picoseconds. For longer-pulse lasers still less than about 100 ps, thicker layers of plasma are heated, and the time to decouple this reservoir of heat from the surface is longer, leading to greater heat-impact in the target. For pulses of roughly 100 fs or less, very little residual heat is left in the material; in biological tissues there may be very little histological impact or inflammatory response, just a few cells away from a laser-cut wound-edge, and thin shock waves will erode as they propagate relatively short distances, due to viscous dissipation.

2.3 Scale lengths for erosion of shock waves

The scaling of ultrashort-pulse lasers leads to another inversion in relationship: for the propagation of laser-created shockwaves. Early in the exponentiation of laser applications, it was identified that strong heating and cavitation from laser interaction could drive shockwaves through water [19]. The nature of such shockwaves is well known: a material surface is driven at a speed greater than the local speed of sound. Pressure increases locally, as the moving surface accrues material faster than it can unload as a pressure wave (acoustic wave). The driving surface ultimately drops to a speed below the speed of sound, and the local region of increased pressure detaches as a propagating feature with a steep front. The steep front propagates at a speed determined from the amplitude of the shock—a speed greater than the speed of sound in the background material.

Originating from a laser focus, the shock wave diverges in 3D, increasing its surface area, and spreading its energy thinner. At some distance depending on the shock strength, the pressure jump degrades until it goes merely sonic.

The transitions at the leading and trailing edges of the shock involve sharp shear stress of the flow, which can result in material or tissue and cell damage. Because of the large shear rates there, the transition is subject to viscous dissipation, with wave kinetic energy converted to local heating, and in the process eroding the shock. Thus the shock weakens and can vanish by viscous effects, in addition to dissipating as the shock diverges and is spread out.

From this arises an additional scaling that ultrashort-pulse lasers can exploit: shockwaves launched in the material from femtosecond pulses are thin, as their driving is brief, and may vanish from viscous dissipation long before their strength is otherwise naturally spread out and reduced. Shocks driven by a fixed fluence of 9 $J\ cm^{-2}$ in corneal tissue show a striking difference depending on mode of fluence-delivery: the range of shocks driven by 150 fs laser pulses is an order of magnitude less than when delivered using 60 ps pulses [6, 7].

2.4 Longer-time diffusion of heat and range of heat-affected zone

The mediating plasma layer decouples rapidly from the substrate, for ultrashort-pulse laser ablation—essentially the absorbed energy is carried away in the ablated material itself, quenching the process. What residual heat has managed to couple to the substrate, in the brief time before quenching, then diffuses normally, but its timescale relation is essentially lost. The residual heat-affected range in this case is determined by the maximum extent of heating in the context of the relevant threshold (e.g., latent heat of melting), as the heating spatially spreads and decays in intensity.

By way of heuristic illustration, consider **Figure 2**, which supposes heat deposited initially in a Gaussian pattern in one dimension, into a homogeneous material. This profile diffuses as a Gaussian distribution, its width increasing and its amplitude decreasing, preserving the total area under the curve (i.e., without energy loss). At any time, a heat-affected zone can be identified (arrows in **Figure 2**) as the range over which the local specific heat (*cf.* temperature profile) exceeds the threshold for a specific effect under consideration, such as melting, or meets the conditions for protein denaturation.[1] In the figures, initially this radius increases, as the heat spreads; but the amplitude also decays, and in the trade-off the boundaries (arrow tips) where the threshold condition is met subsequently shrinks. The ultimate extent is then the maximum extent for which the threshold was met, at any given time.

For single ultrashort pulses, this residual heat is quite small and may not show any heat-affected zone. But for burst-mode delivery of fluence, discussed below, the accumulation of many small residuals can have a useful and controllable impact. The heuristic notion of **Figure 2** will be revisited quantitatively and in simulation later in this discussion.

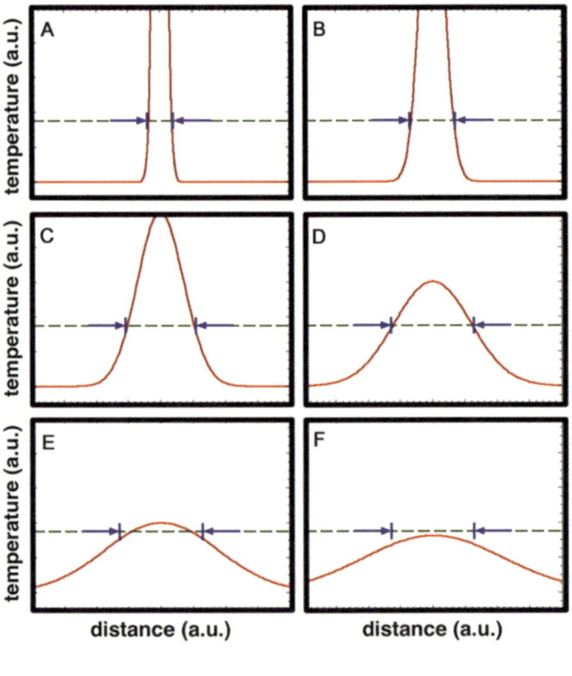

Figure 2.
A schematic depiction of the diffusion of a generic Gaussian temperature distribution over space, for different times (A–F). Where a specific-energy threshold effect (dashed blue line: e.g., necrosis, latent heat of vaporization), provides the reference, neither the initial full-width at half-max (FWHM) nor the thermal-diffusion length-scale L_{th} is the figure of merit for the range of heat impact. Instead, the extent of impact grows, then diminishes, and the ultimate heat-affected zone is the maximum value. Video 1 animation of this can be viewed at https://youtu.be/ioogrzE_46I [8].

[1] In medical applications, this effect is referred to as selective photothermolysis, originally developed by Anderson and Parrish [20] to thermally heat high-contrast targets, such as tattoo ink or superficial blood vessels, in a manner that surrounding tissues are not heated beyond their threshold temperature.

3. Physics relationships of burst-mode ultrashort-pulse material processing

For single ultrashort-pulse laser treatments, at repetition rates much less than about 1 MHz, all plasma and most heat in the substrate will dissipate before the next pulse of a sequence arrives, and the mediating plasma must be re-established by each subsequent pulse. Very little heat is left behind, but it is typically only negligible for repetition-rates less than roughly 100 kHz.

Burst-mode ultrashort-pulses—packets of tens to thousands of pulses delivered at multi-MHz repetition rates—cross time-scales and create new control in two ways: (1) tiny but non-negligible heat left by single ultrashort-pulse lasers will not have dissipated, creating the option for accumulating a *desirable* amount of residual heat, and (2) ionized atoms comprising the plasma may not have fully recombined, meaning that an opportunity is created for simmering the conditions which control plasma-mediated absorption. We introduced burst-mode ultrashort-pulse laser processing as a new mode of fluence-delivery around 1999, [21–24] using amplified trains of 1 ps pulses derived from a feedback-controlled Nd:glass oscillator [25].

In **Figure 1**, red bars illustrate the disparate time scales for pulse duration and for the separation of pulses within a burst—added to this might also be a third bar to indicate the duration of a burst (typically ~100 ns to ~1 μs), and a fourth for the period of the burst-repetition rate. The pulse duration, time-separation between pulses within a burst, total burst-duration, and pause-time between bursts delivered, all have an impact on the amount of heat left within the target, the time-dependent temperature, and the strength of shock waves that may be generated in a biological tissue or material [26–29].

3.1 Slow heat accumulation

As mentioned above, for nanosecond and longer-duration laser pulses, a heat-affected zone (HAZ) surrounding the focal spot may be large, and significant collateral damage may result. Single ultrashort pulses are known typically to leave behind negligible—but not zero—residual heat. Given that burst-mode ultrashort-pulse laser treatments combine both timescales, do they exhibit long-pulse characteristics like the duration of a burst, or ultrashort-pulse characteristics deriving from the individual pulses that comprise the bursts? The answer is that bursts allow controlled accumulation of small amounts of residual heat typical of ultrashort pulse interaction.

To show this, 100 μm-thick aluminum foils were drilled through by single bursts of 1600 1 ps pulses (12 μs @ 133 MHz), at integrated fluences up to 6 kJ cm^{-2}, and hole diameters were recorded (**Figure 3** [8]). As the burst-fluence increases, the size of the hole cut through increases, but not surprisingly this size is not just the imprint of the focal spot—the hole radius is not simply the radius below the Gaussian focal spot distribution at which the specific energy directly imprinted matches the specific latent heat of vaporization. Ablation was not determined by local heat deposition. Instead, the much larger hole sizes seen can be fitted in **Figure 3** by a dependence $E_0^{1/2}$.

The hole diameters produced in aluminum foils, in the series of **Figure 3**, can be matched [8] by finding—over a family of equal-area 2D Gaussians, similar to the heuristic of **Figure 2**—the *maximum* radius for which the specific energy threshold is met, *viz.*,

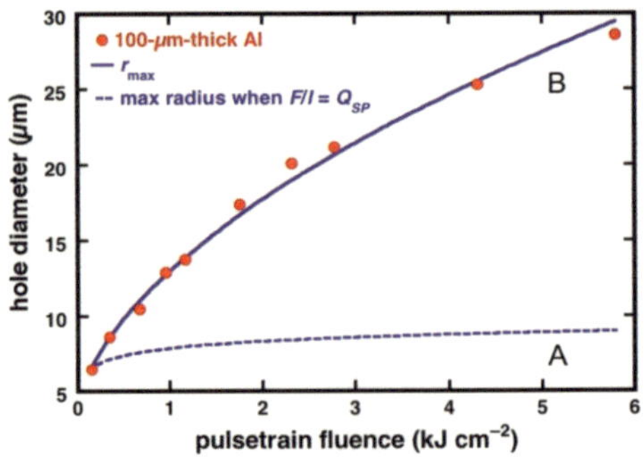

Figure 3.
Measured dependence of through-hole diameter as a function of average irradiant exposure (fluence), for 100 μm-thick aluminum foils under burst-mode processing. The hole size is shown compared to two simple models: (A) the radius within which the Gaussian focal spot locally "prints" sufficient fluence F so that the specific energy F/l equals the latent heat of vaporization; and (B) the maximum extent (cf. ***Figure 2****) achieved starting from an initial Gaussian temperature-profile derived from the laser-spot profile.*

$$r_{\max} = \sqrt{\frac{E_0}{lQ_{sp}e}} \tag{2}$$

with E_0: absorbed energy shared to the foil (after ablative quenching), l: foil thickness (100 μm in **Figure 3**), Q_{sp}: latent heat of vaporization (J cm^{-3}) and e is Euler's number.

Figure 4 illustrates this radius at different times, identifying the maximum above. This shows a model of heat diffusion in aluminum following irradiation by a single 12-μs burst of 1-ps pulses, focused to a spot of 5-μm radius. The simple model is illustrative: it assumes equal-efficiency absorption for all pulses, without hydrodynamics. Two different net-absorbed pulse-energies are compared, corresponding to different pulse-energies or to different absorption efficiencies. The sketched trajectories show the radius *r(t)* at any time of a disk within which the specific energy exceeds the latent heat of melting for aluminum. The two ultimate melt-diameters of **Figure 4** are 15 and 44 μm, to compare to the 30 μm maximum hole-diameter from **Figure 3**.

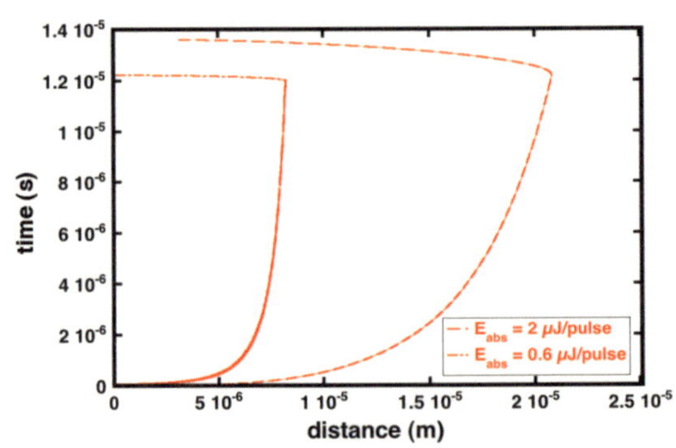

Figure 4.
Model of heat diffusion in aluminum for two possible absorbed laser per-pulse energies. A 12-μs train of 1-ps pulses is focused to a focal-spot radius of 5 μm. The trajectory, for each energy, traces the boundary radius of the disk inside of which the temperature exceeds the melting point of aluminum (cf. heuristic ***Figure 2****, in 1D).*

3.1.1 Ductile cutting of glass

Brittle materials bring their own challenges for laser materials-processing. For 1 ps single-pulse ablation of fused silica, 200 J cm^{-2} was seen to be a limit, corresponding to etch-depths of a few micrometers [8]. Above this single-pulse fluence, the sample typically would shatter in a single shot. Delivering the same fluence, but divided over multiple single pulses at 1 Hz repetition-rate, the weaker shots would cumulatively etch only to about the same ultimate depth before they, too, shattered the glass (**Figure 5** (left)). The final effect was that single pulses delivered at the same location could not ultimately etch more deeply than a few micrometers, regardless of delivering a fixed fluence in a single strong pulse, or divided into multiple weaker shots [21].

This contrasts with the effect of a burst of 300 1 ps pulses at an intraburst rate of 133 MHz, which etch to 10–30 μm in one shot. These etch-depths are deeper than those possible by accumulating single shots. Burst-mode irradiation permitted per-shot fluences of 1–60 kJ cm^{-2}, well beyond the limit at which any number of isolated pulses shattered the fused silica. **Figure 5** illustrates the result: an optically smooth bore, and a lip of material apparently raised while the material was ductile [8].

An important theme of burst-mode ultrashort-pulse materials-processing is that while isolated subpicosecond pulses leave negligible heat behind, this small residual heat can be accumulated within a rapid burst. In practice, glass is transformed from a brittle to a ductile state during a burst, and cutting is achieved during that ductile state.

The accumulated heat results in permanent material changes, as well as this glass-transition softening during processing: after hole-drilling in glass, a change in refractive index can be seen in the zone just outside the channel that has been cut (**Figure 6**).

Figure 7 shows the micro-Raman spectrum of untreated fused silica along with the spectrum sampled for the remelted lip around the ablation crater. The features corresponding to three- and four-member breathing modes show that there is a change in the bond-coordination statistics, which may correspond to densification of the glass. The details of the number of pulses in a burst, their energies, and even the pattern—the distribution of their amplitudes within the burst—offer control over the residual heat accumulated and its range.

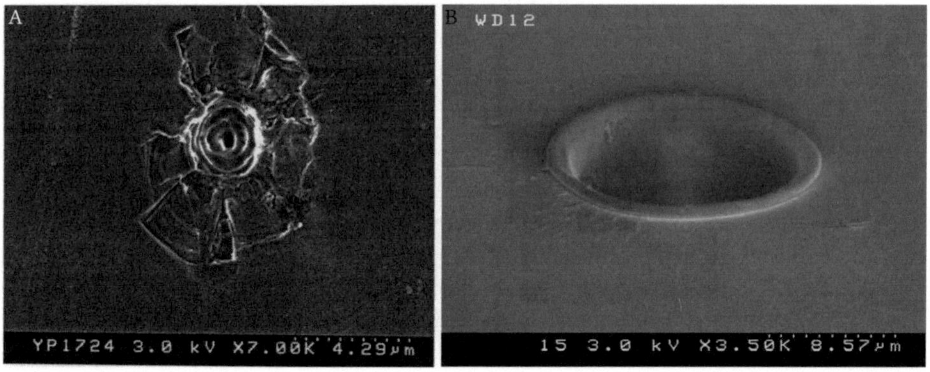

Figure 5.
Scanning electron-microscope (SEM) micrographs of laser-irradiated fused silica. Left: top view, four individual 1-ps pulses at 1 Hz, λ = 1 μm, 93 J cm^{-2} fluence each; Right: oblique view of a ~15-μm deep channel made into BK7 glass by one burst (300 1-ps pulses, λ = 1 μm). The channel has smooth walls and shows no evidence of fractures or cracks.

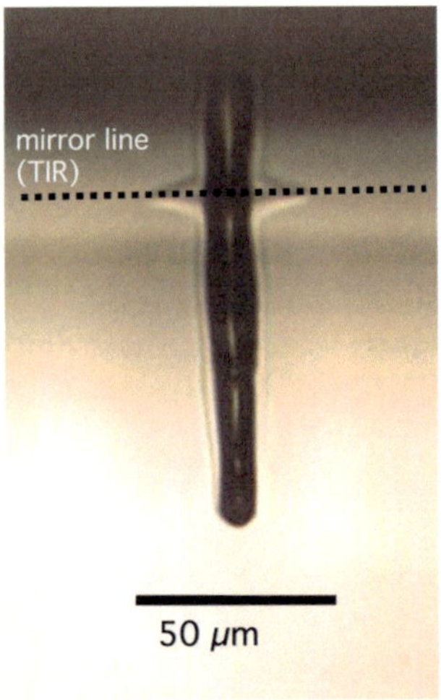

Figure 6.
Hole from one burst-shot drilled in fused silica; optically smooth sides, showing limb of glass with index-of-refraction changes. 300-pulse burst, τ = 1 ps, λ = 1 μm, intraburst rate of 133 MHz. Dotted line indicates top glass-surface making total internal reflection, for this side-view.

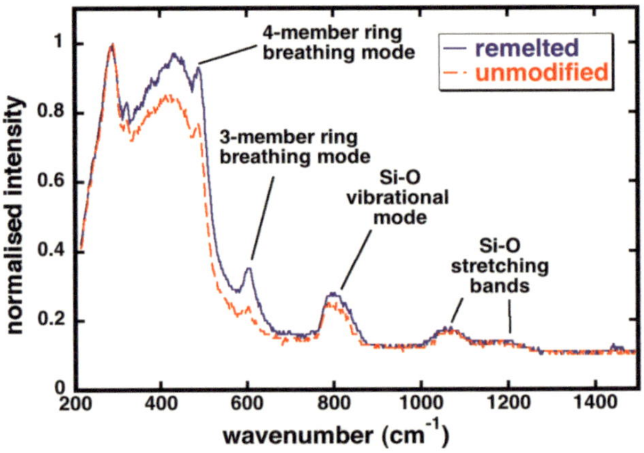

Figure 7.
Micro-Raman spectra (MRS) of fused silica comparing unmodified glass far from the treatment site to the smooth lip of glass around the ablated hole. Where the glass has been melted and resolidified, there are significant changes in the ring-bond structures.

3.1.2 Control of heat accumulation in hard biological tissues

We can demonstrate control of accumulation of small pulse-to-pulse residual heat by simply changing the duration of a burst. **Figure 8** shows burst-mode treatment of *ex vivo* dentin tissue from rat teeth. Trains of about 650 pulses at 133 MHz show thermal changes that are localized, with a kind of melting of the surface of the dentin. Twice that length, about 1300 pulses, results in gross melting and pooling at the lip of the hole drilled. Three times the original length (2000 pulses) exhibits splashing before re-solidification.

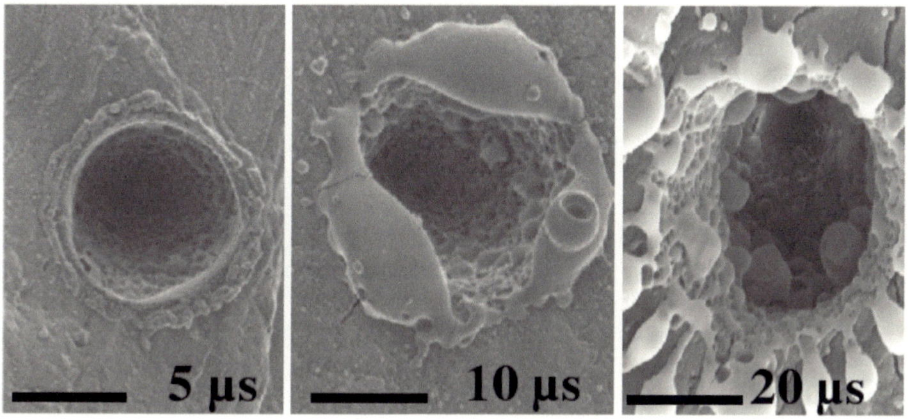

Figure 8.
SEM images of rat dentin tissue irradiated by single bursts of duration 5, 10, and 20 µs (22 kJ cm^{-2}, 44 kJ cm^{-2}, 67 kJ cm^{-2}, respectively). 1 ps pulses at 133 MHz. Magnification scale bars in each image are 5 µm [8, 30].

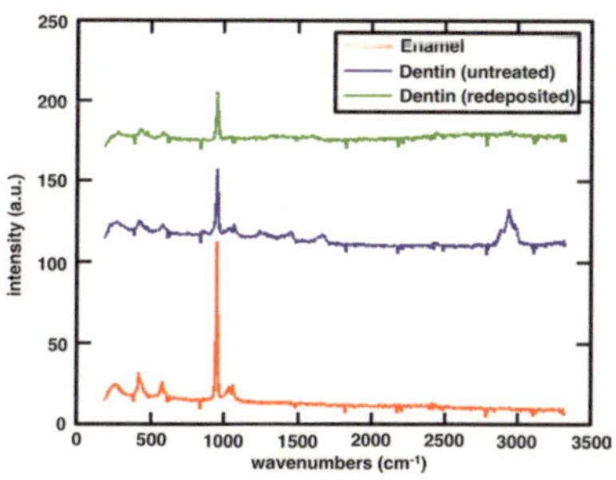

Figure 9.
Micro-Raman spectra (MRS) of native enamel (bottom trace, red) and of both untreated dentin (middle trace, blue) and 1 ps burst-mode treated dentin (top trace, green), intra-burst rate of 133 MHz. The burst-mode laser-modified dentin is glassy in appearance—MRS shows it has almost completely lost its organic components, the dentin converted to a structure more like enamel [8].

Micro-Raman spectra (**Figure 9**) from the inside of laser-cut channels, and from the glassy-looking redeposited material at the hole-lip, show molecular structural changes suggesting that the organic component of dentin ablates away [8, 28].

3.1.3 Incubation effects and division of fluence-delivery

The observed differences between modes of delivery, isolated or rapid bursts, raise the question about how one pulse affects the absorption of the next, when the material or plasma has an effective memory with a characteristic timescale. In laser materials-processing of crystalline and amorphous solids, a well-known effect is "incubation" of crystal defects, subtle material modification, or latent damage [16]. For instance, there is evidence in metals that long-lived crystal dislocations can be accumulated in the bulk, following irradiation with many pulses just below the catastrophic damage threshold. These occur over the depth of heat diffusion, accumulating over many shots before gross damage is manifest. In irradiation of dielectrics at intensities below dielectric breakdown, color centers can be developed. Such subtle preconditioning of the material is thought to have an ultimate impact on ablation damage-thresholds. It might well be expected to be a factor that affects pulse-train burst mode treatments.

For burst-mode ultrashort-pulse interactions, mediated by plasma absorption, we investigated incubation-type effects using a principle of "fluence division"—delivering one, constant, net fluence and constant number of pulses, but dividing it over separate bursts that must restart the interaction each time [8].

This constant total fluence (18 kJ cm^{-2}) was delivered in four different modes: as a single burst of 12 μs, split into two bursts of 6 μs, three bursts of 4 μs, or four bursts of 3 μs, with a delay of a few seconds between bursts to ensure complete relaxation of the processes involved. Thus, a single pulse-train burst of 1600 pulses was partitioned into N smaller sub-bursts, each with n pulses, keeping $N \times n = 1600$. We compared the depth of etching for each of these modes of delivering the same fluence.

Figure 10 shows the results for fused silica; **Figure 11** shows the results for human dentin. In either case, breaking up the burst into sub-bursts created progressively shallower net etch-depths: each time the laser-matter interaction was re-started, there was loss of efficiency.

The loss-trend can be explained by the hypothesis that a certain number of pulses at the start of any burst go into conditioning or incubating the material before ablation really begins, and depth of cutting is proportional to the remainder of pulses in each burst.

This can be interpreted as follows: We posit that for any burst with a given per-pulse fluence a number a of pulses are expended at the start before ablation substantially begins. If one burst consisting of A pulses is partitioned into N mini-bursts, then $N{\cdot}a$ pulses are ultimately lost as "overhead" to seed the ablation process, leaving $A - N{\cdot}a$ pulses to do the etching. If these remaining pulses then etch with unchanged efficiency, once the material has been preconditioned by the initial a pulses each time, then the etch-depth which results will scale as $A - N{\cdot}a$.

For fused silica (**Figure 10**), the scaling fits the data well on condition that a = 155 pulses go into conditioning at the start of any burst. The more times the initial pulse train is interrupted, the greater the overall "overhead" cost.

For dentin, in **Figure 11**, there are two cases with different net fluence: low per-pulse fluence at 0.34 J cm^{-2}, and high fluence an order of magnitude higher at 3.4 J cm^{-2}. For the lower fluence case, a good linear fit suggests a latency equivalent to the first a = 300 pulses of each mini-burst. For the higher-intensity case, a similar fit suggests a latency of about a = 128 pulses. These results may indicate that under these

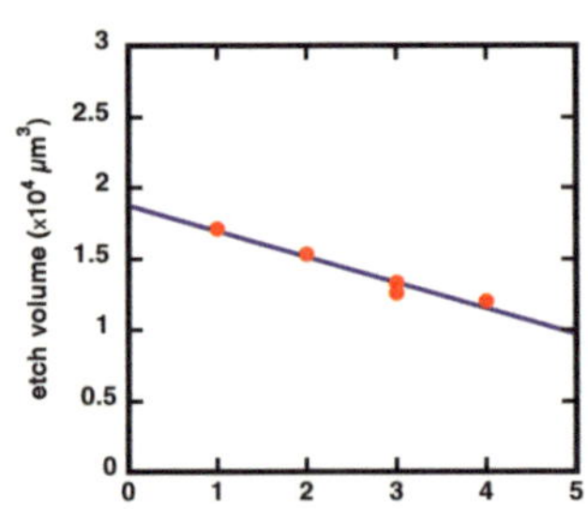

Figure 10.
Fluence-division effect in fused silica: overall fluence, and number of pulses (1600) delivered to each site was kept constant, but delivered in separate bursts: 2 bursts × 800 pulses, 3 × 533, and 4 × 400. "Restarting" the ablation process instead of letting it continue imposes a cost on ablated volume interruption. The "restart cost" is equivalent here to ~155 pulses lost from cutting each time interrupted. Total fluence was fixed at 18 kJ cm^{-2} [8].

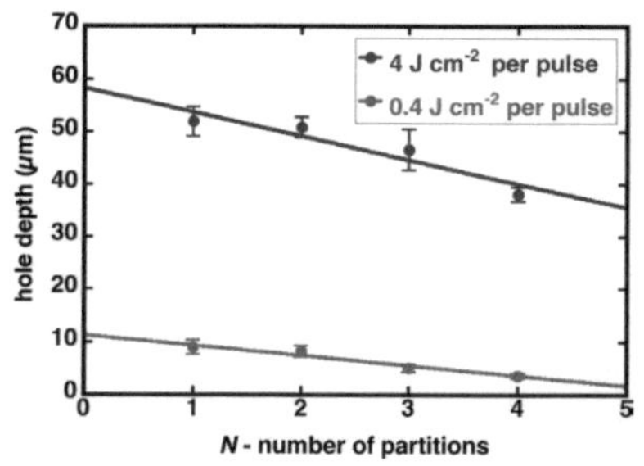

Figure 11.
Fluence-division effect in dentin at two different per-pulse fluences (cf. **Figure 10** *for fused silica). In the lower-fluence case, the "restart-cost" is equivalent here to ~280 lost pulses; this reduces to only ~130 pulses lost, in the case of the higher-fluence interaction. Integrated fluences were kept constant: 5.4 kJ cm^{-2} (lower fluence, green curve) and 54 kJ cm^{-2} (higher fluence, blue curve) [8].*

conditions ablation may not begin immediately, while some transient preconditioning effect is in play, possibly that the density of plasma builds instead of promptly ablating, or perhaps that the material itself changes composition. This is somewhere similar to incubation, but here the process must be reversible, in the sense that restarting the pulse-train burst means that the process must start all over again.

3.1.4 In vitro: range of cell death

Applied to soft tissues and considering the implications for laser surgery, the issue of residual heat from ultrashort-pulse laser ablation in general, and ultrashort-pulse burst treatment in particular, is less about modification of the material left behind and more about survivability for cells remaining in the tissue around the ablated region. We used a standardized tissue-model with the aim of getting consistent and unambiguous results [31]. Hydrogel cell cultures are common soft-tissue phantoms for laser-irradiation [32, 33] and for studies of cell response to drug and radiation treatments (e.g., photodynamic therapy [34] and interstitial laser photocoagulation [35]).

Our soft-tissue model was a 3D agar-based gel phantom: 1% agar content seeded with live F98 rat glioma cells (1–3 × 10^6 cells/mL). Staining protocols included Hoechst 33342, propidium iodide (PI), and Annexin-V (see **Table 1**). After laser-treatment and staining, the samples were 3D-scanned by confocal microscopy, and virtual sections were taken through the ablated regions.

Figure 12 shows the distribution of viable (blue) & necrotic (red) cells from our treated phantom, irradiated by a single burst of 133 pulses of 1 ps duration (1 μs @133 MHz), having a peak pulse-intensity of 4.6 × 10^{13} W/cm^2, a total energy of 1 mJ.

	Live cells	Necrotic	Apoptotic
Hoechst	+	+	+
PI	–	+	–*
Annexin-V	–	+	+

PI can bind to late-apoptotic cells.

Table 1.
Protocols used for staining cells-in-gels phantoms.

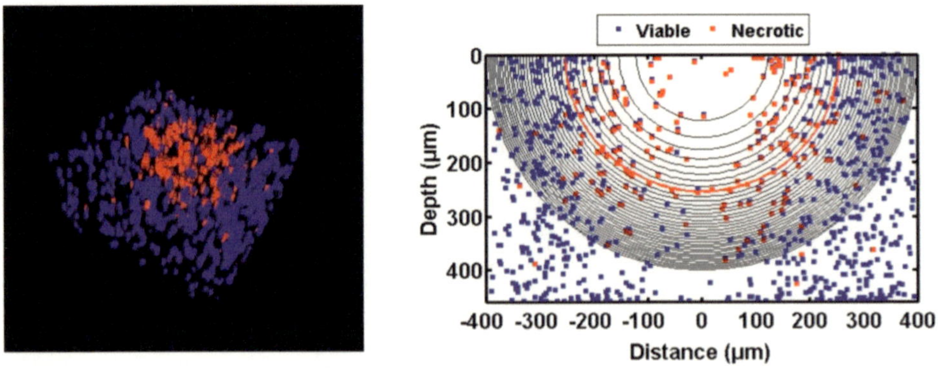

Figure 12.
The 3D distribution of viable (blue) and necrotic (red) cells cultured in hydrogel after irradiation at a peak intensity 4.6 × 10^{13} *W/cm*^{2} *in a 1-µs-duration burst @133 MHz intraburst repetition-rate. Left: 3D confocal-scanning microscope image across the interaction region. Right: the cylindrical projection from 3D of viable and necrotic cells, overlaying the radius-bins used for subsequent analysis. The red solid-line semicircle marks the necrosis range. Video 2 animated 3D rotation can be viewed at https://youtu.be/sA6wmdhtNJg.*

Irradiating over a range of pulse peak intensities 1–5 × 10^{13} W/cm^2, this study showed a necrosis range between 50 and 350 µm that scales as $\sqrt{I}$ where I is the pulse-intensity in the burst (**Figure 13**). This might be understood, other factors being constant, as the absorbed burst-energy spreading out like the surface of a sphere until the specific energy-density drops below a threshold associated with cell necrosis.

The impact above is the result of a single 1 µs burst-shot onto a cultured-cell phantom. In practical use, burst-mode packets are delivered at the interburst repetition rate. Depending on the tissue or material in use, and the system for pointing pulses to different positions across the surgical target, the interburst repetition-rate can also be expected to have an impact on heating.

3.2 Burst-mode ultrafast laser surgery: *ex vivo* articular cartilage

Diffusion of accumulated heat and damage from shock waves in surgery using burst-mode ultrashort-pulse ablation may depend on the tissue type, and even on new effects in complex differentiated tissues of muscle, bone, connective tissue, and nerves. We examined the impact of burst-mode ultrashort-pulse ablation on porcine and ovine articular cartilage.

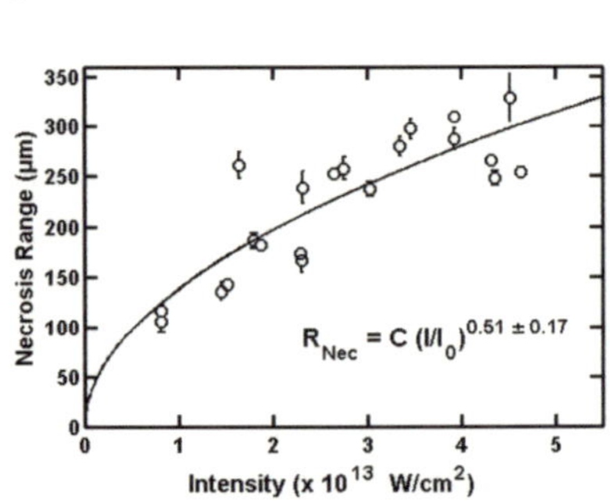

Figure 13.
The cell-necrosis range for the phantom used, as a function of the pulse intensity for a 1-µs-duration pulse-train burst. The fit through the data is a power-law as shown in the figure, where $I_0 = 1.0 \times 10^{13}\,\mathrm{W/cm^2}$*, and* $C = 138 \pm 28\,\mu m$.

The laser for this was a compact Yb:glass fiber, 1030 nm, 300 fs per pulse, 5 ns between pulses (i.e., 200 MHz intraburst rate), 1 ms between bursts (i.e., 1 kHz interburst rate) from FiberLAST Inc./Bilkent University (Turkey). Bursts were 350 ns in duration, comprising about 70 pulses per burst. Average power was up to 240 mW: 3.4 μJ per-pulse energy; 240 μJ per-burst energy.

One question for these studies was about the ultimate depth of cutting, without also changing the laser focus along the axis, deeper into the material—what cutting can be expected under practical conditions? The laser intensity was expected to decrease at the bottom of the channel, for two reasons: loss of propagating laser power to the sidewalls of the channel where plasma would form; and diverging laser spot size, for points past the focal distance.

In previous studies, in which we drilled single channels through metal foils using burst-mode ultrashort-pulse lasers, we characterized a persistent-plasma waveguiding effect. This effect produced channels much longer than a Rayleigh range for the beam [36]. This occurs without translating the laser focus to be deeper into the workpiece, and comes as a result of optically transforming the beam while preserving its global coherence. In order to assess the impact of such guiding in biological tissues, to find the ultimate depth of cutting for laser surgery, and to ascertain what of the effect survives from single channel geometries to extended slices, line-cuts were made in porcine cartilage over multiple passes, without making any changes in focus depth.

Translation speeds 1–10 mm/s were used. Directly after the laser treatment, the cartilage was stained with Calcein AM (live; green fluorescence) and ethidium homodimer (EthD-1) (dead; red fluorescence) to evaluate cell viability[2] and the stained cartilage pieces were sectioned perpendicular to the line-cuts. The live-dead stain fluorescence was imaged using confocal laser-scanning microscopy (CLSM), in a virtual section plane set about 200 μm deep into the tissue (**Figure 14**, Top). Thereafter the tissue was fixed in 10% buffered formalin, paraffin-embedded, and sections stained with hematoxylin and eosin (H&E) (**Figure 14**, Bottom).

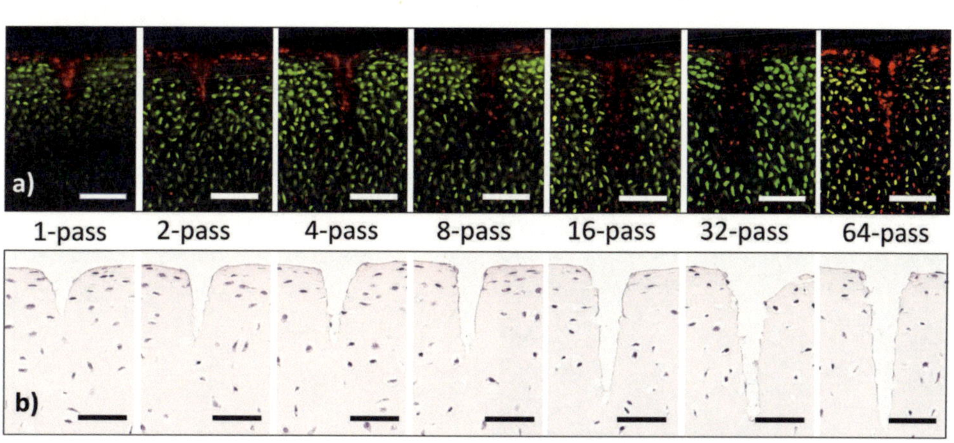

Figure 14.
Line cuts made by repeated passes without repositioning focus, onto porcine articular cartilage, in order to identify ultimate depth of cutting. Translation speed 2.5 mm/s. (a) Live-dead (green-red, respectively) stain was used to identify the range of cell necrosis in the viable tissue away from the cut surfaces. White bars = 100 μm. (b) White-light microscopy of H&E-stained microtome sections. Black bars = 50 μm.

[2] LIVE/DEAD Viability/Cytotoxicity Kit for mammalian cells; Molecular Probes, ThermoFisher Scientific, Mississauga, ON.

The cut-depth without refocusing deeper was seen to saturate after about 16 passes, at a depth of about 200 μm.

3.2.1 From line cuts to graft beds

We then prepared small graft-beds, up to 10 mm wide, cut into ovine cartilage using an *x-y* raster-scan pattern, in this case raising the cartilage toward the lens systematically. A 3-axis high-precision translation stage setup directed the tissue at the laser focus, in lines, in a strictly serial motion, rather than position-dithered. A visible-infrared (IR) spectrograph monitored our depth of cutting through cartilage, employing characteristic spectral lines in the plasma-plume to stop cutting as soon as the ablation touched bone (see **Figure 15**).

The evolving graft-bed was irrigated with phosphate-buffered saline (PBS) medium at intervals during preparation. The prepared graft beds were cultured after laser-cutting, live-dead stained, and then scanned in three dimensions using laser confocal microscopy.

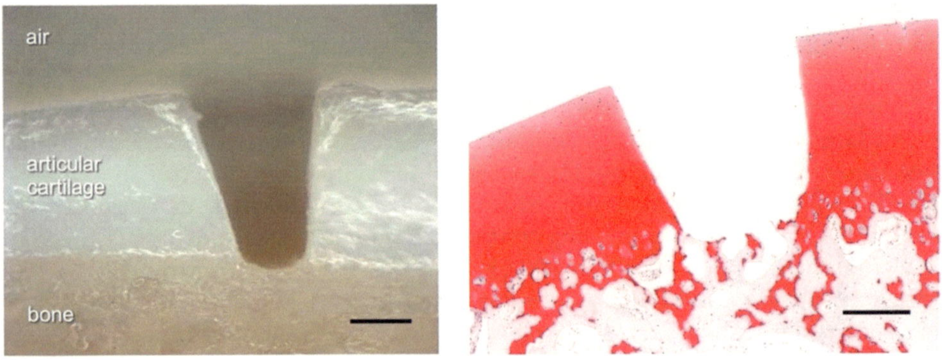

Figure 15.
White light photograph of a simple open-faced section through a small graft bed in ovine articular cartilage (left), showing articular cartilage and underlying bone. Safranin-O stained paraffin section of the graft bed (right). Size-bars 500 μm.

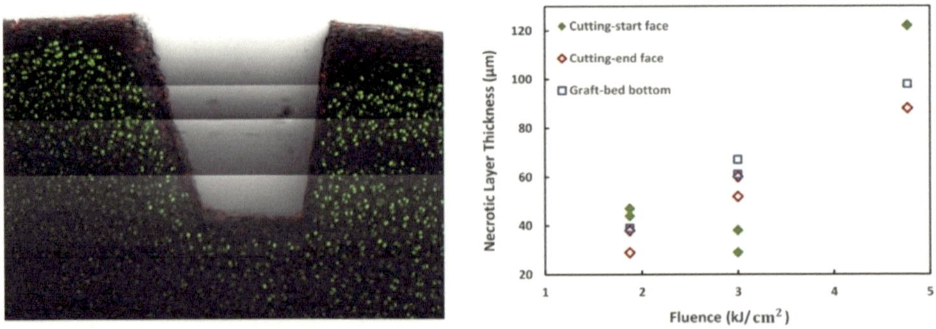

Figure 16.
Laser-scanning confocal imaging provides a virtual section across a small graft bed in articular cartilage. Left: live-dead (green-red, respectively) staining helps to identify the range of dead chondrocytes, in the remaining tissue walls. Right: the range of cell-death can be seen to depend on the overall fluence used to prepare the graft-bed. The bed has an asymmetry associated with raster-scanning: tissue was cut in lines corresponding to in and out of the page, with new lines added progressing left to right as shown in left figure, with the starting side sloped and the ending side more upright. For similar final graft-beds, lower-fluence (gentler) cutting reduces the zone of damaged tissue.

The left image of **Figure 16** is analogous to the left image of **Figure 15**, but is a composite image: it combines a CLSM image of live-dead stain, and a bright-field white-light microscope image of the same sample. The CLSM scan represents a virtual section made for a plane about 200 μm below the section surface.

A small range of cell death is an essential factor in the success of integration of cartilage tissue-grafts, and below we show the range of dead chondrocytes brought about by integrated laser fluences in the range 1–5 kJ cm^{-2}. It's immediately evident how significant an effect it is that most of the heat of burst-mode ultrashort-pulse laser cutting is carried off in the ablated material itself. The range of tissue damage is less than expected from frictional heat in mechanical removals such as drilling and sawing. However, it seems clear as well that the slow serial processing in *x-y* raster scans likely accumulates more heat than is necessary, burst after burst at 1 kHz [37].

4. Persistence of plasma and burst-mode ultrashort-pulse lasers

As discussed above, for single ultrashort pulses at repetition rates much below 1 MHz, all plasma and most heat in the material dissipates before the next pulse, and the material largely returns to its original state. For multi-MHz intraburst rates, neither plasma nor heat dissipates, meaning that an opportunity is created for keeping alive the transient plasma that governs plasma-mediated absorption. This can be both hindrance and help, depending on details. It's a "feature," if one can exploit linear absorption in a residual plasma lying close to the target, meaning that one does not have to re-initiate breakdown. In that case the burst can be shaped, after the first few pulses, to have smaller pulses, less likely to cause strong shocks, optimally tailored to the surgical or processing needs. It's a "bug," if substantial plasma plume develops, and interferes with deposition of laser energy onto or close to the ablating target-surface. In practice, the nature of laser-target interaction for any given pulse depends on the history of *all* previous pulses before it, and possibly in complicated ways.

4.1 Where does the laser energy go?

We studied the development of plasma conditions throughout a burst of pulses by making a complete laser-energy accounting: measuring the transmission, specular reflection, and diffuse backscatter for each of the pulses during a burst-irradiation. Knowing the energies of each incident laser pulse, and the disposition of all light that was *not* absorbed, let us infer the full dynamics of burst-mode absorption [38].

The measurements amounted to a full laser-calorimetry setup (**Figure 17**). A reference of incident laser light was measured pulse by pulse using a beam-sampler and a photodiode. Transmitted laser light was collected immediately below the target using an apertured integrating tube coated internally with barium sulfate high-reflectance coating, and fitted with a high-speed photodiode. Diffuse-backscatter light was collected by a similarly coated integrating sphere suspended above the target and fitted around the lens tube holding the 8 mm aspherical target lens, with a 2 mm entrance hole situated about 100 μm above the target, almost in contact with the fused silica targets. A high-speed photodiode collected the light which filled this integrating sphere. Laser light which was specularly back-reflected from the plasma was re-collected by the target lens, naturally collimated backwards along the beam, and a sample was relayed to the input face of a large-core multimode fiber, with fast photodiode at the output end.

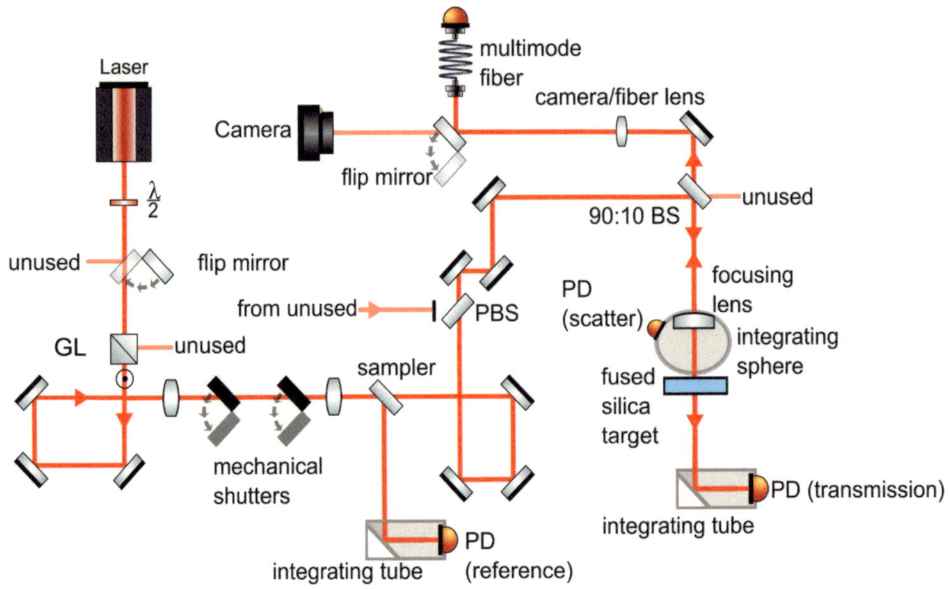

Figure 17.
Schematic of the optical layout for time-resolved laser calorimetry, to infer dynamic absorption. Direction of propagation is indicated by red arrows. PBS: polarizing beam splitter, GL: Glan-laser polarizer, λ/2: half-wave plate. All of the energy from light incident on the fused silica target ends up in one of four places: transmitted, specularly reflected, diffusely backscattered, or absorbed by the plasma.

4.2 Measurement of absorption dynamics

With this full accounting for incident laser light, and all laser light not absorbed, the light lost to absorption in the plasma was inferred, pulse-by-pulse for the entire burst. **Figure 18** shows the dynamic changes in specular reflection, diffuse backscatter, and transmission, along with the computed history of absorption during burst-irradiation.

What this shows: specular reflection is high for about 15 pulses of the burst: each pulse sees the plasma it has itself created afresh during 300 fs, added on top of any plasma that still survives 5 ns after the previous pulse (and longer times for effects of earlier pulses). The initial plasmas each pulse creates is expected to be thin, dense, and sharply defined, but as residual ionization or weak plasma accumulates in small remainders following hydrodynamic expansion from pulses 5 ns, 10 ns, 15 ns earlier, it's expected that the plasma critical-density surface would evolve to be farther out from the substrate surface, thereby reducing specular reflection.

The diffuse backscatter signal is small at all times, indicating an accumulating plume is not very substantial for these conditions and timescales. Given 5 ns each time, to expand, disperse, and recombine before the next pulse arrives, the accumulated plume does not greatly refract or scatter light. Self-transmission of laser pulses is relatively high for the first pulse, and the next pulse or two of a burst, but then shows a sharp drop. This is consistent with the expected formation of a thin plasma overlying the fused silica substrate, with an initially steep density gradient (scale-length less than a laser wavelength) around critical density. In the thin overdense plasma the laser light will be evanescent and not propagate, but will effectively tunnel energy through to the glass, in the manner of frustrated total internal reflection. As the plasma develops over the initial few pulses, its thickness increases and the transmission drops; this is consistent with the fluence-division observations above. Late in the

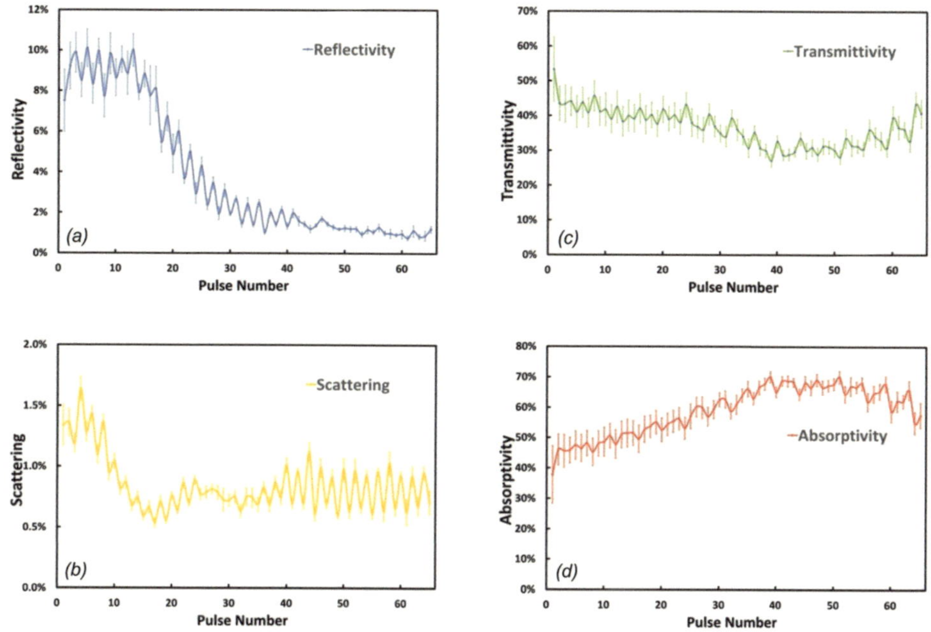

Figure 18.
Dynamics of (a) specular reflection, (b) diffuse backscatter, (c) transmissivity, for each pulse within a burst. Knowing each input pulse energy, a full accounting of laser light not recollected gives the time-dependent absorption (d). Laser: λ = 1030 nm, 300 fs pulses @ 200 MHz; 60 pulses/burst [38].

burst, the transmission begins to rise again, which may indicate that the accumulation of plasma has degraded the ability of focused pulses to reach the substrate and form a reflective plasma, or possibly an extended plasma has refracted the incident light enough to cause it to pass outside the nominal focal spot.

By complete accounting of all laser light *not* absorbed, we find the absorption dynamics from first to last pulse of the burst. We can see the jump in absorption from about 30% for the first pulse, in its context as a single isolated pulse on a dielectric, to more than 40% as the mediating plasma is established, and after that a steady increase in absorption throughout the burst, reaching 70% before declining slightly. This late decline of absorption results from the rise in transmission instead.

4.3 Picosecond probing of plasma persistence

From the time-resolved absorption of each pulse incident on the target, we can infer much about the plasma and plume that gradually builds over the whole time of the burst. A finer-grained understanding of what happens to the plasma created by each pulse, in the 5 ns period *between* pulses, would lay out the role of the persistence of the plasma from the end of one pulse until it becomes the initial condition for the next pulse.

Figure 19 sets out our schematic for creating a burst-mode ultrashort laser-pulse produced plasma, and then probing that plasma precisely, at times ranging from just before and up to 300 ps after each pulse, for every pulse in the burst. The laser is polarization-divided into pump and probe beams, and the probe is delayed in a timing-slide to arrive at a selected time in the range −30 ps to +5 ns relative to the pump pulses on target. The rapidly dissipating plasma created by each pump-pulse throughout the whole burst is probed this way, so we can examine whether the

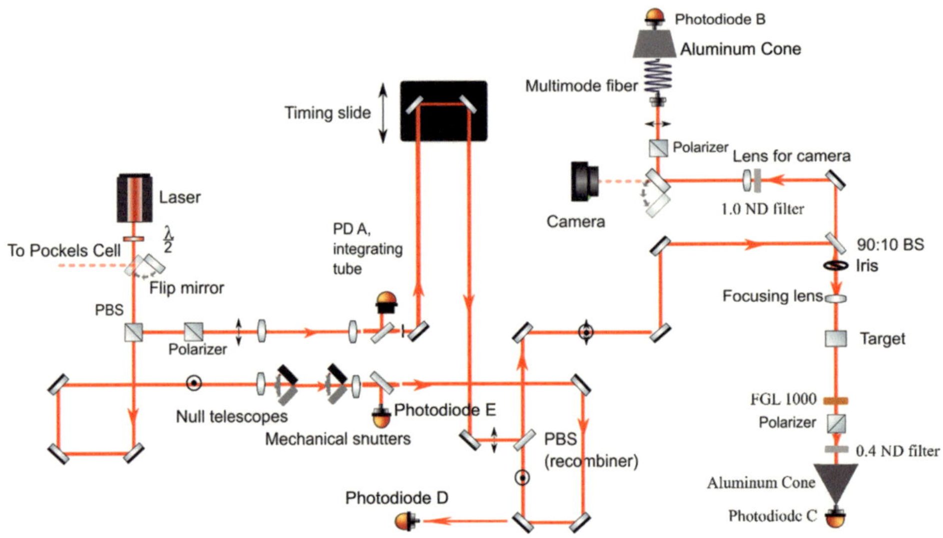

Figure 19.
Schematic of the optical setup for pump-probe studies of specular reflectivity and transmissivity for times up to 300 ps following each pulse of a burst of 60 pulses. Polarization is indicated with black arrows for horizontal polarization (parallel to optical table), and black dotted-circles for vertical polarization. Direction of beam travel is indicated by red arrows. PBS: polarizing beam splitter, GL: Glan-laser polarizer, λ/2: half-wave plate.

dynamics of plasma creation and persistence are the same for early pulses, when the fused silica target is bare, as they are for pulses late in the burst, when "fresh" plasma is being added on top of some manner of accumulated background.

Pump and probe beams are propagated collinearly to the target, and timed to each other using frequency-domain interferometry. Two aspects of the plasma are probed: specular reflectivity and beam transmissivity. Because the probe is distinguished from the pump by its orthogonal polarization, diffuse backscatter or scattered transmission that do not preserve the laser beam k-vector cannot be directly measured. It was verified that polarization cross-talk signal from the pump line into the probe channel line was less than 5%.

4.3.1 Plasma persistence: specular reflection and transmission

Figure 20 summarizes the specular reflectivity and direct transmissivity of the probe from the plasmas created by the pump pulses, for times from 30 ps before each of the pump pulses arrives until 300 ps after them. This 330 ps time-dependence relaxation is summarized in separate plots for the 3rd, 11th, and 19th pulses in the burst.

The Fresnel reflectivity of bare fused silica is a few percent for laser pulses, if no plasma is created. For each pulse n = 3, 11, 19 in the burst, there is some specular reflectivity seen for the probe for times before the pump pulse has arrived—of course, these are times about 5 ns *after* at least one previous pump pulse on target. The reflectivity seen 30 ps before pulse 19 has arrived is half the reflectivity of flat native fused silica, but by the time of the 19th pulse, nearly 100 ns into the burst, a crater at the surface has already been created, and reduced reflectivity is expected.

For the first 50 ps after each pump-pulse of a burst has ended, the probe sees specular reflectivity *growing*. This reflects that the plasma is an initially very thin layer, much of it at a density above critical density. Within this thin overdense plasma

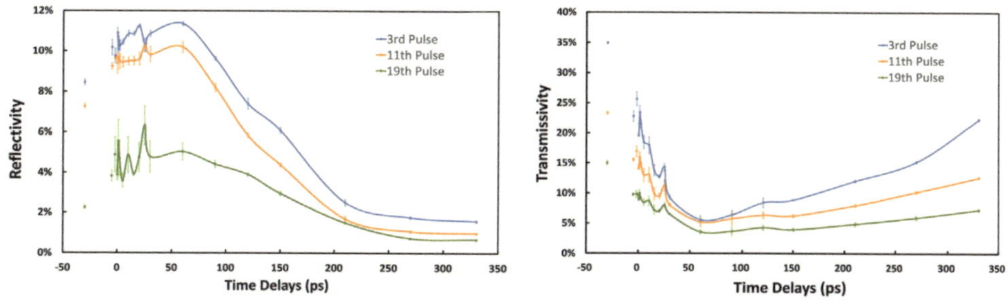

Figure 20.
Summarizes the evolution over 300 ps of the probed specular reflectivity and probed direct transmission of the plasmas created by each of the pump pulses. The evolution is similar but not identical, for the 3rd, 11th, and 19th pulses in the burst [38]. (NB: the single data point at −30 ps stands alone, the connecting line does not imply a smooth temporal evolution up to the time of arrival of the pump pulse).

the wave has an evanescent solution. The light tunnels, and recovers a real olution on the other side of the overdense plasma, as discussed above in the context of (self-) transmitted light in the absorption measurements of the pump, above.

As the plasma layer begins to expand over tens of picoseconds, and the gradient scale-length increases while the plasma is still overdense, the reflectivity can rise, and the transmissivity will drop. As expansion continues, past 100 ps, the unloading plasma will eventually go underdense, at which point reflectivity will drop and probe-transmissivity will again increase. Over nanoseconds the plasma dissipates, absorptivity decreases, and eventually reflectivity and transmissivity return to values nearer to native fused silica, albeit with a cratered surface scattering both the reflected and transmitted light, and likely with a little plasma ionization remaining.

The effect is most pronounced for pulses early in the burst, here the 3rd pulse. For pulses late in the train, like the 19th, the plasma that has slowly accumulated during the burst degrades the contrast of reflection and transmission, as does ablative damage to the surface.

5. Conclusions

Intense ultrashort-pulse lasers that employ plasma-mediated absorption minimize residual heat—not because of a diffusion timescale, but through a competition between heat diffusion inward, and the dissipation of hot plasma in which the laser pulse has deposited energy. This "ablative quenching" is distinct from the mechanisms governing heat transfer in nanosecond laser heating.

Burst-mode ultrashort-pulse lasers add additional timescales—the timing between pulses within a burst, and the duration of the entire burst—which depend on new characteristics of the interaction physics, and therefore provide additional measures of control of the heat left behind in the surgical tissue or material being processed. Rapid repetition of pulses, here at 133 MHz or at 200 MHz intraburst rates, leads to the persistence of plasma from one pulse to the next. Keeping the plasma "alive" gives a universal and more predictable and universal linear absorption, without repeating dielectric breakdown, and even without the linear absorption typical of native chromophores, which themselves may vary from location to location in a sample.

This persistence of linear absorption, and not having to re-initiate dielectric breakdown, also means that the opportunity is created to use lower-intensity pulses subsequent to the first few of a burst, which affords the possibility of reducing shockwave damage. Thus the duration of a burst, intra-burst pulse-spacing, and indeed the pattern of intensities of pulses within a burst, all together determine the impact, and support different modes of fluence-delivery customized to particular applications [26, 27, 39].

Acknowledgements

This work was accomplished with funding from the Natural Sciences and Engineering Research Council of Canada, under the Discovery Grants Program (RGPIN-2017-06757), and the Strategic Partnership Grants for Projects (STPGP 494025-16).

Particular thanks for the hands-on contributions of summer-students and co-op students Soho Shim, Lee Sikstrom, Luke Coulter, Amber King, and Sara Early.

The 300 fs burst-mode laser was designed, constructed, and/or improved by Hamit Kalaycioglu, Seydi Yavas, Sohret Görkem Karamuk, variously of the University of Bilkent, and FiberLAST laser company, Türkiye, and currently at Lumos Laser A.Ş., Türkiye. Can Kerse made important contributions in establishing the laser and experimental setup. Professor F. Ömar Ilday from the University of Bilkent, Turkey, and currently at Ruhr Universität Bochum (RUB), Germany, made invaluable and essential contributions to laser design and experimental principles.

Author details

Robin S. Marjoribanks[1*], Jinseng Tang[1], Thomas Dzelzainis[1,2], Melissa Prickaerts[3], Lothar Lilge[3], Margarete Akens[3,4], Colin Veevers[1], Nick N. Gharabaghi[1], Andreas Hitzler[1], Seydi Yavas[5] and Sohret Görkem Karamuk[5]

1 Department of Physics, University of Toronto, Canada

2 Rutherford-Appleton Laboratory, UK

3 Department of Medical Biophysics, University of Toronto, Canada

4 TECHNA Institute, University Health Network, Canada

5 Lumos Laser A.Ş., Turkey

*Address all correspondence to: marj@physics.utoronto.ca

References

[1] Hecht J. Beam: The Race to Make the Laser. Oxford, England, UK: Oxford University Press; 2005

[2] Liu X, Du D, Mourou G. Laser ablation and micromachining with ultrashort laser pulses. IEEE Journal of Quantum Electronics. 1997;**33**(10):1706-1716

[3] Mourou GA et al. Method for Controlling Configuration of Laser Induced Breakdown and Ablation, 5,656,186. 1997

[4] Du D, Liu X, Mourou G. Reduction of multi-photon ionization in dielectrics due to collisions. Applied Physics B: Lasers and Optics. 1996;**63**(6):617-621. DOI: 10.1007/BF01831002

[5] Du D, Liu X, Korn G, Squier J, Mourou G. Laser-induced breakdown by impact ionization in SiO_2 with pulse widths from 7 ns to 150 fs. Applied Physics Letters. 1994;**64**(23):3071-3073. DOI: 10.1063/1.111350

[6] Juhasz T, Kastis GA, Suárez C, Bor Z, Bron WE. Time-resolved observations of shock waves and cavitation bubbles generated by femtosecond laser pulses in corneal tissue and water. Lasers in Surgery and Medicine. 1996;**19**(1):23-31. DOI: 10.1002/(SICI)1096-9101(1996)19:1<23::AID-LSM4>3.0.CO;2-S

[7] Juhasz T, Loesel FH, Kurtz RM, Horvath C, Bille JF, Mourou G. Corneal refractive surgery with femtosecond lasers. IEEE Journal of Selected Topics in Quantum Electronics. 1999;**5**(4):902-910

[8] Marjoribanks RS et al. Ablation and thermal effects in treatment of hard and soft materials and biotissues using ultrafast-laser pulse-train bursts. Photonics & Lasers in Medicine. 2012;**1**(3):155-169. DOI: 10.1515/plm-2012-0020

[9] Stuart B, Feit MD, Rubenchik AM, Shore B, Perry MD. Laser-induced damage in dielectrics with nanosecond to subpicosecond pulses. Physical Review Letters. 1995;**74**(12):2248-2251

[10] Joglekar AP, Liu H, Spooner GJ, Meyhöfer E, Mourou G, Hunt AJ. A study of the deterministic character of optical damage by femtosecond laser pulses and applications to nanomachining. Applied Physics B: Lasers and Optics. 2003;**77**(1):25-30. DOI: 10.1007/s00340-003-1246-z

[11] Pronko PP et al. Avalanche ionization and dielectric breakdown in silicon with ultrafast laser pulses. Physical Review B. 1998;**58**(5):2387-2390

[12] Pronko PP, Dutta SK, Squier J, Rudd JV, Du D, Mourou G. Machining of sub-micron holes using a femtosecond laser at 800 nm. Optics Communication. 1995;**114**:106-116

[13] Bettis JR, House RA II, Guenther AH. Laser induced damage in optical materials. In: Proceedings of the 8th Annual Symposium on Optical Materials for High Power Lasers; Gaithersburg, MD. Washington, DC, USA: Optica; 1976. pp. 338-345. DOI: 10.6028/NBS.SP.462

[14] Wood RM. Laser Damage in Optical Materials. Boston: Adam Hilger; 1986

[15] Campbell JH, Rainer F, Kozlowski MR, Wolfe CR, Thomas IM, Milanovich FP. Damage resistant optics for a megajoule solid state laser. In: Bennett HE, Chase LL, Guenther AH,

Newnam BE, Soileau MJ, editors. SPIE Proc. Vol. 1441 - Laser-Induced Damage in Optical Materials: 1990. Boulder, CO: SPIE; 1991. pp. 444-456. DOI: 10.1117/12.57232

[16] Guo B, Sun J, Hua Y, Zhan N, Jia J, Chu K. Femtosecond laser micro/nano-manufacturing: Theories, measurements, methods, and applications. Nanomanufacturing And Metrology. 2020;**3**(1):26-67. DOI: 10.1007/s41871-020-00056-5

[17] Chowdhury IH, Xu X. Heat transfer in femtosecond laser processing of metal. Numerical Heat Transfer, Part A. 2003;**44**:219-232. DOI: 10.1080/10407780390210224

[18] Kerse C et al. Ablation-cooled material removal with ultrafast bursts of pulses. Nature. 2016;**537**(7618):84-88. DOI: 10.1038/nature18619

[19] Bell CE, Landt JA. Laser-induced high-pressure shock waves in water. Applied Physics Letters. 1967;**10**(2):46-48. DOI: 10.1063/1.1754840

[20] Anderson RR, Parrish JA. Selective photothermolysis: Precise microsurgery by selective absorption of pulsed radiation. Science (1979). 1983;**220**(4596):524-527. DOI: 10.1126/science.6836297

[21] Herman PR, Oettl A, Chen KP, Marjoribanks RS. Laser micromachining of 'transparent' fused silica with 1-ps pulses and pulse trains. In: Proceedings of SPIE Conference on Commercial and Biomedical Applications of Ultrafast Lasers. Vol. 3616. Bellingham, WA, USA: SPIE; 1999. pp. 148-155

[22] Herman PR et al. Laser micromachining of transparent glasses and aluminum with ps-pulse bursts at 1054 nm. In: Pacific Rim Conference on Lasers and Electro-Optics, CLEO - Technical Digest. New York, NY, USA: IEEE; 2000. DOI: 10.1109/cleo.2000.907417

[23] Herman PR, Marjoribanks RS, Oettl A, Chen K, Konovalov I, Ness S. Laser shaping of photonic materials: Deep-ultraviolet and ultrafast lasers. Applied Surface Science. 2000;**154**:577-586. DOI: 10.1016/S0169-4332(99)00463-8

[24] Herman P, Marjoribanks R, Oettl A, Burst-Ultrafast Laser Machining Method, US 6,552,301 B2 [Online]. 2003. Available from: https://patents.google.com/patent/US6552301B2/en [Accessed: February 6, 2024]

[25] Marjoribanks RS, Budnik FW, Zhao L, Kulcssr G, Stanier M, Mihaychuk J. High-contrast terawatt chirped-pulse-amplification laser that uses a 1-ps Nd:glass oscillator. Optics Letters. 1993;**18**(5):361-363

[26] Lickschat P, Demba A, Weissmantel S. Ablation of steel using picosecond laser pulses in burst mode. Applied Physics A: Materials Science & Processing. 2017;**123**(2):1-10. Article 137. DOI: 10.1007/s00339-016-0743-y

[27] Balage P et al. Advances in femtosecond laser GHz-burst drilling of glasses: Influence of burst shape and duration. Micromachines (Basel). 2023;**14**(6):1-10. Article 1158. DOI: 10.3390/mi14061158

[28] Kerse C, Kalaycioılu H, Elahi P, Akçaalan Ö, Ilday FÖ. 3.5-GHz intra-burst repetition rate ultrafast Yb-doped fiber laser. Optics Communication. 2016;**366**:404-409. DOI: 10.1016/j.optcom.2015.12.064

[29] Neuenschwander B, Kramer T, Lauer B, Jaeggi B. Burst mode with

ps- and fs-pulses: Influence on the removal rate, surface quality, and heat accumulation. In: Laser Applications in Microelectronic and Optoelectronic Manufacturing (LAMOM) XX. Washington, DC, USA: SPIE; 2015. p. 93500U. DOI: 10.1117/12.2076455

[30] Dille C, Kaifosh P, Forrester P, Mordovanakis AG, Lilge L, Marjoribanks R. Ablation of hard dental tissue using ultrashort pulsetrain-burst (>100MHz) laser. In: 2009 Conference on Lasers and Electro-Optics and 2009 Conference on Quantum Electronics and Laser Science Conference, CLEO/QELS 2009. Washington, DC, USA: Optica; 2009

[31] Qian Z et al. Pulsetrain-burst mode, ultrafast-laser interactions with 3D viable cell cultures as a model for soft biological tissues. Biomedical Optics Express. 2014;**5**(1):208. DOI: 10.1364/boe.5.000208

[32] Pérez-Gutiérrez FG, Camacho-López S, Aguilar G. Time-resolved study of the mechanical response of tissue phantoms to nanosecond laser pulses. Journal of Biomedical Optics. 2011;**16**(11):1-9. Article 115001. DOI: 10.1117/1.3644380

[33] Cherian AV, Rau KR. Pulsed-laser-induced damage in rat corneas: Time-resolved imaging of physical effects and acute biological response. Journal of Biomedical Optics. 2008;**13**(2):1-10. Article 024009. DOI: 10.1117/1.2907214

[34] Ali MFM. Topical delivery and photodynamic evaluation of a multivesicular liposomal Rose Bengal. Lasers in Medical Science. 2011;**26**(2):267-275. DOI: 10.1007/s10103-010-0859-9

[35] Sramek C et al. Improving the therapeutic window of retinal photocoagulation by spatial and temporal modulation of the laser beam. Journal of Biomedical Optics. 2011;**16**(2):028004. DOI: 10.1117/1.3542045

[36] Dean J et al. The effects of degraded spatial coherence on ultrafast-laser channel etching. Optics Express. 2008;**16**(18):13606-13616

[37] Prickaerts MJ. Biological Effects of Ultrafast Burst-Mode Laser Ablation of Chondral Tissues [MSc thesis]. Toronto, ON, Canada: University of Toronto; 2019

[38] Marjoribanks RS et al. Plasma persistence, accumulated absorption, and scattering: How physics lets us control the heat left behind in ultrafast-pulse burst-mode laser surgery. In: SPIE Proc. Frontiers in Ultrafast Optics: Biomedical, Scientific, and Industrial Applications XXIV. San Francisco: SPIE; 2024

[39] Desbiens L et al. Flexible and programmable pulse shaping MOPA fiber laser platform, performances and applications. The Review of Laser Engineering. 2013;**41**(9):691. DOI: 10.2184/lsj.41.9_691

Section 2

Laser for Material Analysis

Chapter 6

Modernization and Automation of Gemological Testing: Harnessing the Power of Laser-Induced Spectroscopy for Raman, Photoluminescence, and Photoluminescence Lifetime Analysis

Wenxing Xu

Abstract

The chapter explores the revolutionary application of laser-induced spectroscopy in the field of gemstone analysis. It highlights the use of advanced techniques such as Raman spectroscopy, photoluminescence, and photoluminescence lifetime analysis, which have proven instrumental in identifying a variety of gemstones such as corundum, spinel, emerald, alexandrite, etc. This state-of-the-art technology stands at the forefront of differentiating natural from lab-grown materials, discerning treatment methods, and determining the origins of significant colored gemstones. This chapter provides an in-depth view of the practical implementation of emission spectroscopy in the analysis of colored gemstones, shedding light on its integral role in the rapidly evolving gemological testing systems. The goal is to furnish readers with a thorough understanding of how these advanced methods are transforming gemstone analysis, contributing significantly to the industry's quest for enhanced precision and accuracy.

Keywords: gemological testing, non-distractive testing, characterization of gemstones, Raman spectroscopy, photoluminescence, PL lifetime analysis

1. Introduction

In the realm of gemological science, the quest for methodologies that are precise, non-destructive, efficient, and straightforward for the identification of gemstone materials presents a significant and ongoing challenge. This challenge is particularly pronounced in the context of the rapidly advancing technologies in gemstone synthesis and treatment. Recent decades have underscored the need for reliable identification tools within the gemstone industry. This need is primarily driven by the prevalence of melee diamonds in the jewelry market. Despite their relatively low individual value, these diamonds carry substantial implications for brand reputation, especially in cases where the presence of laboratory-grown or treated diamonds is not transparently disclosed [1–4]. Additionally, the surge in popularity of colored diamonds, notably pink diamonds, which are often the product of complex multi-treatment processes, adds another layer of intricacy to the task of gemstone identification [5–12].

In light of these challenges, there has been a notable acceleration in the development of advanced spectroscopic methods, such as photoluminescence [13–17] and photoluminescence lifetime analysis (PLLA) [18–20]. These methodologies are complemented by the advent of new, portable gemstone testing devices, including Raman spectrometers (e.g., [21, 22]) and long-wave ultraviolet (LWUV) fluorescence spectrometers (e.g., [23]), which have rapidly evolved in the realm of diamond testing. Of particular note, the applicability of these methods extends beyond diamonds and is increasingly being recognized in the broader field of colored gemstone testing.

In the controlled settings of gemstone laboratories, a suite of established gemological testing methods is employed to rigorously identify and characterize gemstones. The use of microscopy allows for the detailed examination of both internal and external features of gemstones, utilizing binocular microscopes. Refractometry, a technique that measures the refractive index of gemstones, plays a critical role in their identification. Likewise, specific gravity testing, which assesses the density of a gemstone relative to an equal volume of water, is a key determinant in identification processes. Beyond that, ultraviolet testing, examining gemstones' responses to ultraviolet light, forms another crucial component of their identification, belonging to the classic testing methods in gemology. Additionally, spectroscopic methods, encompassing fourier-transform infrared spectroscopy (FTIR), UV-VIS, and Raman spectroscopy, probe the interaction between different types of light and the gemstones, yielding insights into their complex physical and chemical compositions, which are mainly useful for solving questions concerning the identification of varieties and for treatment examination. Furthermore, while sparingly used, destructive testing methods, such as acid testing, can be crucial in the identification of certain challenging gemstones [24]. In modern gemological laboratories, LA-ICP-MS plays a crucial role, enabling precise analysis of trace elements in gemstones. This aids in their identification and authentication. By detecting elements at very low concentrations, LA-ICP-MS assists gemologists in assessing the origins and authenticity of gemstones, thereby contributing significantly to the quality control process in the industry [25].

This chapter delves into the pivotal role of gemological laboratories in maintaining transparency within the industry, focusing on the application of spectroscopic techniques, including Raman spectroscopy, photoluminescence (PL), and photoluminescence lifetime analysis (PLLA). These advanced methods have proven highly effective in differentiating between natural and lab-grown gemstones, identifying various treatment methods, and potentially ascertaining the origins of significant colored gemstones. Furthermore, the discussion included herein aims to provide a comprehensive and detailed overview of the state-of-the-art methodologies used in gemological testing, highlighting their significance in an industry where precision, innovation, and transparency are paramount.

2. Portable Raman spectroscopy

In the field of gemology, the ability to accurately classify mineral types is the cornerstone of gemstone identification and valuation. Traditional methods of gemstone assessment, reliant on visual inspection and basic physical properties, often fall short in the face of diverse and overlapping features of minerals [26–29]. Sophisticated systems, such as micro-Raman systems, offer more definitive identification but come with high costs and complexities, limiting their use to well-equipped laboratories.

Against the backdrop, portable Raman spectroscopy emerged as a game-changer in the gemstone industry. Unlike the traditional, more complex micro-Raman systems, portable Raman devices offer a balance of cost-effectiveness, ease of use, and speed, without compromising the accuracy of mineral identification [22, 30–33]. These devices leverage the unique Raman scattering features of gemstones and are complemented by automatic spectral analysis algorithms, enabling rapid, accurate, and simple mineral-type classification. This advancement not only democratizes the use of advanced technology in smaller-scale gemological labs and retail settings but also enhances the ability to test a wide variety of gemstone samples quickly and efficiently in standard office conditions.

Take the portable Raman spectroscopy device described by Tsai & Xu [22] as an example, which is equipped with fiber-based 405 nm laser spectroscopy. The device's design simplifies sample alignment, meets safety standards, and enables it to be operated effectively in routine office environments, making it an ideal tool for identifying a broad range of gemstones. The device has been proven to have a broad mineral testing capability, and it is effective in differentiating between natural and synthetic gemstones, isolating mineral subcategories, and identifying rare gemstones. In the subsequent section, a selection of case studies is presented that effectively highlight the remarkable accuracy, efficiency, and clarity of the portable 405 nm laser Raman equipment, underscoring its utility and effectiveness in routine gemological practices.

2.1 Distinguishing between gemstones and their simulants

For the gemologist, distinguishing gemstones from their imitations is a crucial task. Common imitations of diamonds in the gem market include colorless corundum, topaz, cubic zirconia (CZ), gadolinium gallium garnet (GGG), and yttrium aluminum garnet (YAG). Each of these materials has characteristic Raman scattering spectra, which are key to their identification. Diamonds are identified by a single peak at 1337.5 cm^{-1}, indicative of the sp^3 diamond lattice [34]. Corundum displays intense peaks at 424 and 758 cm^{-1}, and topaz is characterized by a distinct peak at 937 cm^{-1}, primarily due to Si–O vibration [35]. Cubic zirconia shows a strong, broad band at 622 cm^{-1}. GGG and YAG, both synthetic materials with high refractive indices and that are used as diamond simulants, have distinct peaks at 746, 598, and 361 cm^{-1} (for GGG) and 787, 727, and 269 cm^{-1} (for YAG), enabling their differentiation from diamonds. These Raman spectral features are essential for accurately separating these gemstones from their imitations.

However, both diamonds and their simulants typically exhibit a high refractive index, often surpassing the upper limit of traditional refractometers. Consequently, differentiating between them necessitates the use of multiple gemological testing techniques, including specific gravity testing and the use of handheld spectroscopes, among others. In this context, Raman spectroscopy stands out for its remarkable efficiency, offering a clear advantage in rapidly and accurately distinguishing between diamonds and their various imitations (**Figure 1**).

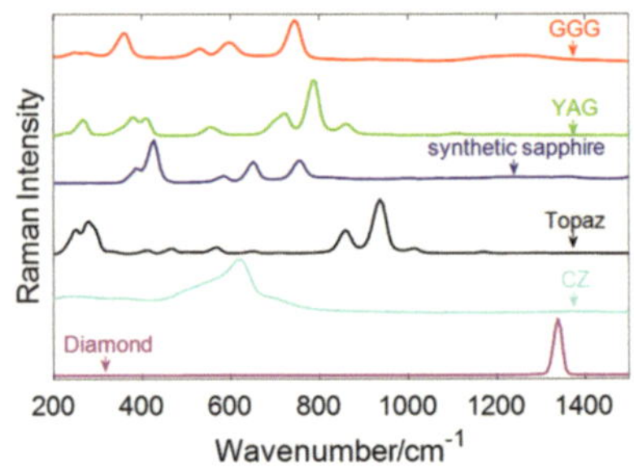

Figure 1.
Typical Raman scattering spectra for diamond and commonly encountered diamond simulants, including GGG, YAG, colorless sapphire, topaz, CZ, and diamond; source [22].

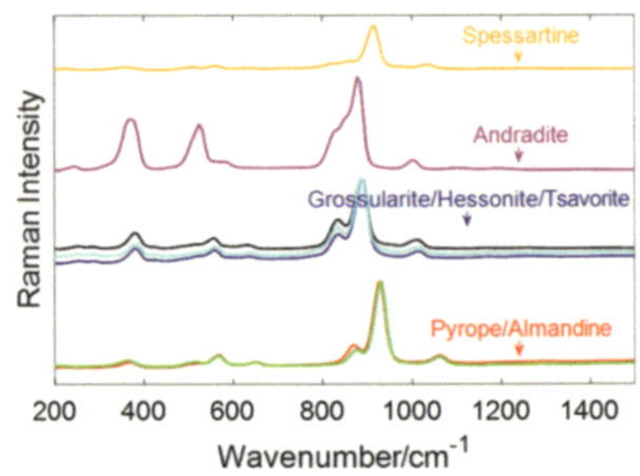

Figure 2.
Raman scattering spectra for garnet varieties, source [22].

2.2 Distinguishing between varieties in the same mineral group

Mineral groups such as garnet, tourmaline, and feldspars have distinct varieties impacting gemstone quality, identifiable by their Raman spectra when using a 405 nm laser based on chemical composition. For instance, the 405 nm Raman spectra of various garnet species (pyrope-almandine-spessartine, grossularite, hessonite, tsavorite, and andradite) show unique patterns. The pyrope-almandine-spessartine series displays similar spectra with intense Si—O vibrations, while the grossularite group and andradite garnet each have distinct spectral characteristics. This allows for differentiation between key garnet varieties without needing further chemical analysis (**Figure 2**).

2.3 Distinguishing between natural and lab-grown gemstones

When it comes to distinguishing between natural and lab-grown gemstones, the portable Raman spectroscope distinctly identifies variations in the Raman spectra of synthetic spinel compared to natural spinel. Natural and flux spinels exhibit characteristic peaks at 322, 412, 671, and 772 cm^{-1}. However, flux spinels show broader peaks, signaling their disordered structure owing to higher manufacturing temperatures (900–1200°C) than the temperatures that produce spinel in nature [36]. In contrast, Verneuil synthetic spinels, which are formed at higher temperatures (2150–2250°C), present a unique spectrum with significantly broader peaks at 343, 426, 527, 697, 792, and 872 cm^{-1}, indicative of an even more disordered structure (**Figure 3**).

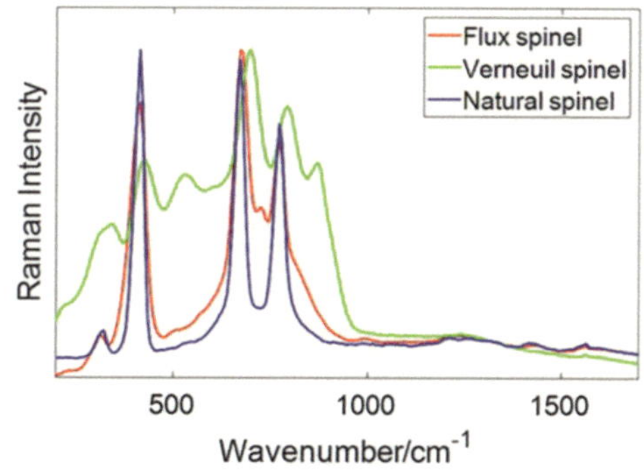

Figure 3.
Raman scattering spectra for natural spinel, source [22].

These distinctions are consistently observed in over 80 Gemological Institute of America (GIA)'s collections. Verneuil spinel features a higher refractive index (RI) of 1.728 compared to the natural spinel's RI of 1.718. Meanwhile, flux spinels have the same RI as natural spinels, meaning additional tests such as photoluminescence and LA-ICPMS are commonly needed for differentiation. Raman spectroscopy simplifies the process by providing a clear and singular method for distinguishing these variants, thus significantly enhancing efficiency in gemological testing.

The 405 nm laser Raman spectroscopy setup, covering a range of 150–4500 cm^{-1}, also effectively characterizes natural and synthetic emeralds, differentiating them based on the presence of hydroxyl vibrations at around 3500 cm^{-1}. Furthermore, flux synthetic emeralds lack the water-related peaks present in natural and hydrothermal synthetic emeralds, while Biron hydrothermal synthetics are uniquely identifiable by a peak at 2822 cm^{-1} [22]. These factors are relevant when it comes to differentiating between low-iron emeralds, such as those from Colombia, and high-iron emeralds from Russia and Zambia, which 405 nm Raman spectroscopy can facilitate. Colombian emeralds show only type 1 water peaks at 3615 cm^{-1}, whereas Russian and Zambian emeralds exhibit both type 1 and type 2 water-related peaks at 3605 and 3666 cm^{-1} [37]. Thus, the 405 nm Raman spectroscope can effectively identify various types of synthetic and natural emeralds, including high-iron natural emeralds and different hydrothermal synthetics (**Figure 4**).

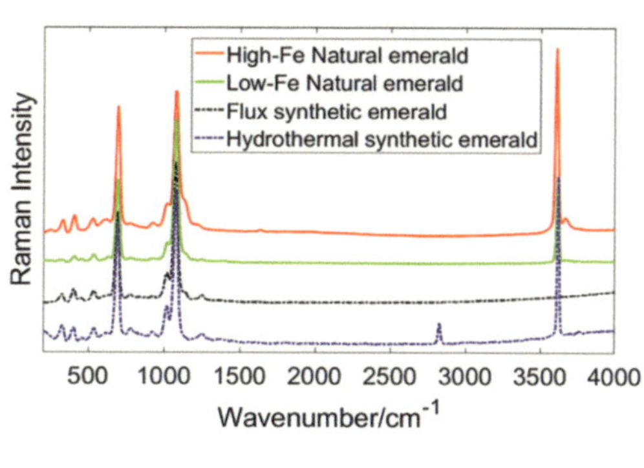

Figure 4.
Raman scattering spectra for high-Fe natural, low-Fe natural, flux synthetic, and hydrothermal synthetic emerald; source [22].

2.4 Discussion

The common laser wavelengths used in portable Raman systems are 405, 532, 785, and 1064 nm [22, 38–40]. While a confocal Raman system, for example, Renishaw products such as the inVia Reflex and Virsa Raman analyzer, commonly uses a 532 nm and also includes 660 and 786 nm lasers for excitation [41, 42]. These wavelengths are widely utilized for their efficiency in Raman spectroscopy applications. The 405 and 532 nm laser are known for its high Raman signal efficiency, ideal for inorganic materials and mostly used in gemstone analysis; the 785 nm laser is popular due to its effectiveness for over 90% of Raman active materials with limited interference from fluorescence; and the 1064 nm laser is better suited to colored and darker materials such as natural products, dyes, oils, and colored polymers [39].

The use of a 405 nm laser, for example, as the excitation source in portable Raman spectroscopy, offers several distinct advantages over other excitation light sources in Raman analysis. Firstly, it provides a flat background baseline from wavenumbers 100 to 4000 cm^{-1}, minimizing the need for background correction. This feature simplifies the analysis process significantly. Secondly, it delivers a strong and clear signal at around 3500 cm^{-1} area, making it highly effective for studying crystal water in gemstones. The presence of hydroxyl groups in crystal water, commonly found in gemstones such as the beryl group, tourmaline, and topaz, provides valuable information for their identification. Traditionally, hydroxyl groups are analyzed using FTIR in gemological labs, particularly for emeralds, where FTIR testing is essential to distinguish natural from flux synthetic and Biron hydrothermal synthetic varieties [43–45]. However, with the pronounced hydroxyl signal offered by the 405 nm portable Raman system, the FTIR test may be reduced, thereby streamlining the identification process in gemological practices. In contrast, the Raman spectrum collected using a 532 nm laser as the excitation source has its hydroxyl signal obscured by the photoluminescence spectrum between 600 and 800 nm, leading to a loss of completeness in the information.

The effectiveness of the 405 nm laser in Raman spectroscopy largely stems from its optimal choice of excitation wavelength, which significantly influences the interplay between Raman and photoluminescence spectra. Raman spectroscopy, which measures changes in energy due to light interacting with molecular vibrations, reports values relative to the excitation source as a Raman shift in cm^{-1} [46]. In contrast, photoluminescence sheds light on the composition and structure of materials, with a spectral range extending from 400 to 1000 nm when using a Raman spectrometer with a charge coupled device (CCD) detector [47]. Selecting an excitation wavelength that minimizes photoluminescence is key to reducing noise and enhancing the signal-to-noise ratio in Raman spectra [48]. Furthermore, this careful selection helps avoid overwhelming fluorescence backgrounds, which can hinder accurate Raman measurements. Therefore, the relationship between Raman spectroscopy and photoluminescence underscores the critical role of excitation wavelength choice in achieving precise and reliable results in both techniques [46].

Nonetheless, the Raman spectroscopy probe system, while effective in many scenarios, does have certain limitations in its sensing capabilities. For instance, its spectral resolution restricts its ability to differentiate fine variations in the Raman spectra of certain subcategories. Additionally, samples with strong blue background fluorescence, such as some natural diamonds, and translucent minerals, such as turquoise and lapis lazuli, may encounter interference from background fluorescence. Moreover, in the case of rubies with high chromium content, the Raman scattering can be absorbed by the chromium, leading to a reduction in signal strength. Additionally, Raman spectroscopy is not always sufficient for distinguishing between gemstones of the same mineral type that differ in trace element concentrations, as seen in ruby and sapphire, both corundum varieties or in emerald and alexandrite, where chromatic element impurities influence their color. These subtle variations typically do not result in notable changes in the Raman spectra captured by this device [22].

3. Innovative photoluminescence analysis

Photoluminescence analysis is a non-destructive analytical technique used in gemology to identify treated and synthetic gemstones. It involves illuminating a material, often with a laser, and recording the resulting luminescence as a plot of emitted light intensity versus wavelength. This technique is essential for identifying optically active defects in diamonds, which can indicate treatments or synthetic origins. It is particularly important for type II diamonds (both colorless and fancy-color) and colorless type IaB diamonds [15]. The technique is also used to study the spatial distributions of luminescent properties in gem materials, providing valuable insights for gemologists and scientists. Photoluminescence spectroscopy is, therefore, a vital tool for major gemological laboratories in differentiating between natural and treated gemstones [16] and various mineral materials [49].

3.1 Identification of varieties

In the world of color gemstones, especially among the top five gemstones, namely ruby, sapphire, emerald, spinel, and alexandrite, chromium plays a pivotal role in the coloration of various gemstones, imparting distinctive and vibrant hues. In emeralds, chromium is the key element responsible for their rich green color, creating an intense and vivid hue that is characteristic of this gemstone. Similarly, in rubies, chromium acts as the chromophore, and its incorporation into the aluminum oxide crystal structure produces the gem deep red color. Furthermore, in the realm of sapphires, particularly pink and red varieties, chromium is the color-influencing factor. Likewise, alexandrite, a unique color-changing variety of chrysoberyl, is also colored by chromium. This element is critical for the remarkable phenomenon where alexandrite displays different colors under different lighting conditions—green in daylight and red under incandescent light [27, 50, 51]. Additionally, chromium is responsible for the coloration of certain semi-gemstones, such as the vivid green tsavorite garnet or chromium-colored green tourmaline, which derives its striking color from chromium and is sought after in the gemstone market. This illustrates the significant role played by chromium in defining various gemstone varieties.

Figure 5 presents a comprehensive view of the typical chromium photoluminescence (PL) spectra observed in several significant colored gemstones. These include corundum, spinel, emerald, alexandrite, and zoisite, each demonstrating a distinct chromium PL spectrum signature. Furthermore, gemstones such as corundum, spinel, and chrysoberyl often incorporate chromium into their solid solution matrix [43], generally resulting in a distinct chromium luminescence spectrum. Additionally, red beryl is highlighted, showcasing its characteristic manganese (Mn) PL spectrum, unique to its crystal structure. The comparison between the emerald and red beryl is particularly noteworthy, as they exhibit markedly different PL spectra. This contrast

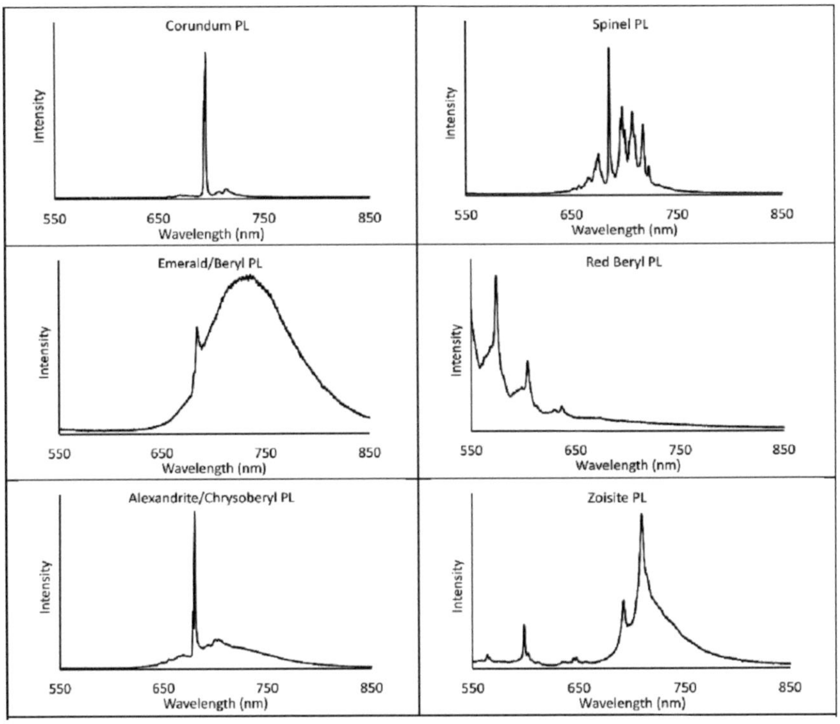

Figure 5.
Typical photoluminescence (PL) spectra of corundum, spinel, emerald, red beryl, alexandrite, and zoisite.

exemplifies how PL spectroscopy can be utilized to differentiate between various varieties within the same species.

To further delve into photoluminescence analysis, we explore its critical role in identifying the presence of chromium in gemstones, thereby aiding in the classification of their varieties. When applied to green beryl minerals, Raman spectroscopy can detect chromium in the fluorescence spectrum, categorizing the mineral as an emerald. Conversely, the absence of chromium-induced fluorescence in beryl indicates that it is not an emerald but rather green beryl or aquamarine, with its coloration attributed to iron impurities. The same principle applies to green tourmaline; the presence of chromium in its fluorescence spectrum identifies it as chromium-tourmaline, whereas the absence of chromium suggests it is an iron-colored tourmaline. This approach underscores the significance of PL spectroscopy in the precise identification and categorization of gemstones based on their compositional and structural characteristics.

In summary, photoluminescence analysis serves as a complementary spectroscopic method to Raman spectroscopy, offering additional insights crucial for gemstone identification. While Raman spectroscopy excels in determining the mineral species, photoluminescence analysis goes a step further. It not only characterizes the mineral based on, for example, chromium luminescence spectra but also sheds light on other chromophoric elements and emission-capable trace elements within the gemstones [49]. This dual capability is invaluable for accurately identifying gemstone varieties and effectively compensates for any limitations in Raman spectroscopy.

Additionally, in professional gemological laboratories, the use of UV-visible spectroscopy (UV-VIS) is a standard practice in the study of the coloration origin. The integration of photoluminescence analysis into this process introduces an additional, robust technique. This advancement in gemological tools enhances the precision and depth of analysis, allowing for a more comprehensive understanding of a gemstone's composition and characteristics.

3.2 Identification of treatment

3.2.1 Heat treatment

Photoluminescence (PL) analysis has been instrumental in studying the heat treatment of natural spinel. The results of photoluminescence spectroscopy have shown that heat-treated spinel exhibits distinct changes in its luminescent properties, which can be used to differentiate heated from unheated spinel. Additionally, PL spectroscopy has been used to analyze the effects of low-temperature heat treatment on blue to violet spinel, revealing broader absorption bands and changes in chromium emission peaks. These findings demonstrate the utility of photoluminescence analysis in identifying heat-treated spinel and understanding the impact of heat treatment on its optical properties [36, 52].

To investigate the effects of heat treatment on natural spinel, a series of controlled thermal experiments were conducted, primarily within the lower temperature range of 600 to 800°C. The objective was to meticulously observe the characteristic photoluminescence (PL) at each temperature level, with a keen focus on discerning any subtle changes. As demonstrated in **Figure 6** of the experiment results, minimal color enhancement was detected at these lower temperatures. However, a slight modification became perceptible when the temperature reached the thresholds of 750 to 800°C, indicating the beginning of the so-called order–disorder behavior in the crystal structure, when an Mg cation exchanges its site with an Al cation [53–58]. As such, this careful application of photoluminescence analysis provided valuable insights into the thermal behavior of natural spinel and its potential for color alteration through

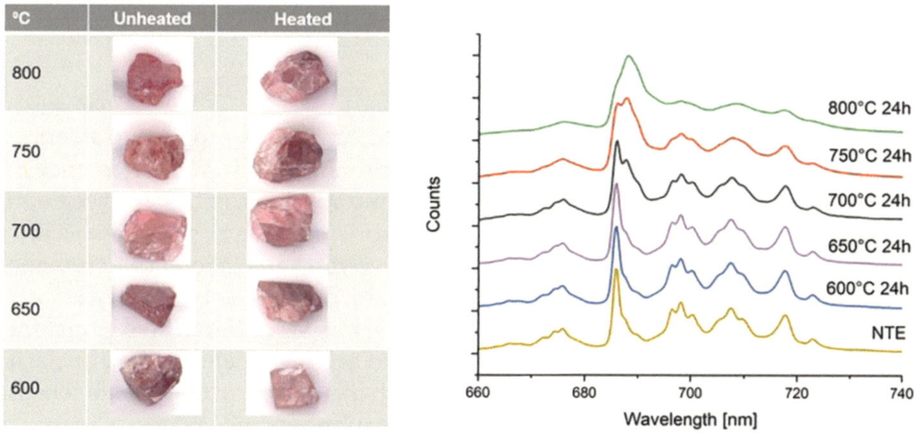

Figure 6.
The outcomes of heat treatment experiments on natural spinel over 24 hours, conducted at temperatures of 600°C, 650°C, 700°C, 750°C, and 800°C (left). And the corresponding changes observed in the photoluminescence (PL) spectrum of the spinel due to these varying temperatures (right).

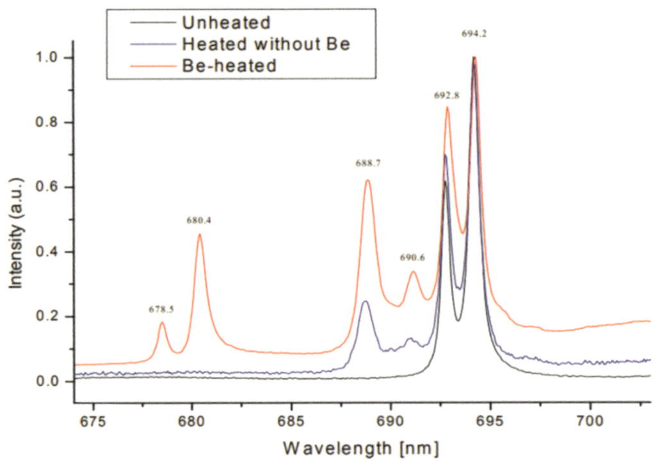

Figure 7.
Luminescence spectra of unheated sapphire (black; showing only the emission of Cr^{3+} in the corundum structure), heated at 1650°C (blue; Cr^{3+} also in the boehmite structure), and Be-treated sapphire (red; Cr^{3+} also in the chrysoberyl structure) (source [14]).

heat treatment.Furthermore, in several recent investigations, photoluminescence (PL) analysis has been utilized to detect heat and beryllium treatments in natural sapphires—treatments that are commonly applied to enhance the gemstones' color and clarity. The PL spectra of sapphires subjected to heat treatment reveal significant alterations, notably the emergence of new Cr^{3+} luminescent peaks that signify modifications within the corundum structure (**Figure 7**). More precisely, sapphires treated with beryllium exhibit luminescent peaks that are distinct and characteristic of the chrysoberyl structure, differing markedly from the emission spectrum of unaltered corundum [13, 14]. This research underscores the critical nature of these spectral changes as a diagnostic tool for the detection of beryllium, offering a non-destructive and reliable method for tracing the treatment history of gemstones. While the LA-ICP-MS technique has long been the standard for identifying beryllium, PL analysis provides a sophisticated and less invasive alternative.

3.2.2 Clarity enhancement

Photoluminescence analysis has emerged as a technique also applied to assess clarity treatments in gemstones. This method uses multi-excitation fluorescence imaging to detect treatment materials. By employing specific excitation wavelengths, this technique can highlight the fillers used in treatments, such as oils or resins in emeralds, while minimizing interference from the gemstone's inherent fluorescence. The innovative approach employs a combination of LED light sources and a color camera with filters to capture detailed fluorescence images of the treated gemstones. This research showcases the potential for rapid and accurate identification of clarity enhancements, particularly demonstrated through the detection of common emerald fillers, thereby streamlining the gemological evaluation process [59].

3.3 Application of PL in origin determination

Photoluminescence (PL) analysis is a valuable tool in the pre-selection and origin determination of gemstone, for example, corundum. The influence of chromium on luminescence is significant, but it is not the only factor; iron also plays a crucial role in modulating fluorescence intensity. The presence of iron is known to quench fluorescence in many minerals. This suggests that the levels of iron can significantly influence the luminescent properties of gemstones. This phenomenon is particularly important in gemstone identification, where the presence or absence of luminescence can provide insights into a gem's composition and inherent characteristics. Notably, higher concentrations of iron in gemstones are often associated with reduced luminescence. Understanding this relationship is crucial for interpreting their luminescent behavior [16, 17, 60].

In the specific context of rubies and sapphires, there is a notable contrast in their iron (Fe) and chromium (Cr) contents. Rubies are characterized by a high chromium concentration (at least 2000 ppm, reaching up to 10,000 ppm) and variable iron levels (ranging from undetectable to several thousand ppm), as noted in various studies (see [61]). Conversely, sapphires are predominantly composed of high iron levels (a few hundred ppm to tens of thousands of ppm) and low chromium contents (ranging from undetectable to around 1000 ppm) (see [62]). Consequently, the fluorescence intensity in rubies is typically much more pronounced than that in sapphires. This variation in intensity can be semi-quantitatively analyzed either through spectrum intensity or by measuring integration time at a fixed luminescence level, which is especially useful given that a ruby's luminescence often surpasses the spectrometer's upper limit. **Figure 8** illustrates the correlation between the Cr/Fe ratio's chemical characteristics and the integration time needed for a fixed luminescence intensity. The scale for integration time spans from microseconds (μs) to seconds (s), offering a comprehensive spectrum to distinguish between rubies and sapphires. For instance, a marble-type ruby, known for its low iron content, requires only a few hundred microseconds of integration time. In contrast, a basaltic sapphire, with its high iron content, necessitates up to several seconds to achieve a luminescence intensity comparable to that of the ruby.

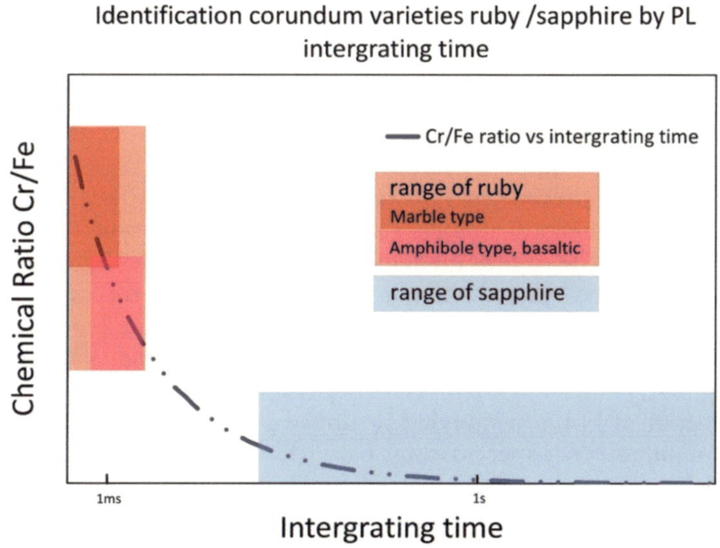

Figure 8.
Correlation between the Cr/Fe ratio's chemical characteristics and the PL integration time.

Moreover, natural rubies can be differentiated based on their iron contents. Marble-type rubies, typically containing less than 1000 ppm of iron and originally from countries such as Myanmar, Vietnam, and Tajikistan, exhibit significantly stronger PL compared to high-iron amphibole type and basaltic rubies (iron contents over 2000 ppm) from locations such as Mozambique and Thailand [61]. This distinction in PL intensity between marble- and high-iron-type rubies further underscores the utility of PL analysis in not only identifying gemstone varieties but also in tracing their geographical origins.

3.4 Discussion

Photoluminescence analysis represents a significant advancement in gemological diagnostics, potentially revolutionizing traditional ultraviolet testing methods. Ultraviolet testing, while a subjective observational technique, is prevalent in daily gemological practice. For instance, assessing the fluorescence strength under long-wave and short-wave UV light is a key method for distinguishing between marble and amphibole-type rubies. Additionally, the detection of chalky fluorescence under short-wave UV light serves as a critical indicator of heat treatment. Moreover, long-wave UV handheld lights are commonly utilized to evaluate the presence of clarity enhancement fillers within gemstone fissures. Now, the introduction of portable photoluminescence spectrometers transforms these subjective assessments into quantifiable visual spectra. This shift not only dramatically increases the sensitivity of detection but also aids in drawing more objective conclusions. Moreover, it contributes significantly to the digitization of gemological testing processes, heralding a new era of precision and standardization in the field.

However, there are several important limitations to note: First, a significant number of gemstone varieties do not exhibit luminescence phenomena. Second, in certain cases, fluorescence spectra can appear similar. For example, the fluorescence spectra of chromium in silicate minerals often show a high degree of similarity, leading to only minimal differences in the fluorescence spectra between emeralds and chromium-bearing tourmaline. In such situations, the assistance of Raman spectroscopy is necessary to accurately identify the specific species or varieties of these materials.

4. Photoluminescence lifetime analysis

Photoluminescence lifetime analysis (PLLA) has become an increasingly valuable tool in the field of gemological studies. This advanced form of spectroscopy detects the temporal events in the environment of a fluorophore, particularly focusing on the decay types signified by a reduction in photoluminescence following excitation (**Figure 8**). Such techniques have found success in the analysis of biomolecular structures [18], nitrogen-vacancy defects in diamonds [19], and rapid testing of diamond melee sorting [63], evidencing their versatility (**Figure 9**).

The application of PLLA in studying Cr-doped gem materials, specifically using 405 nm photoluminescence spectroscopy, has yielded impressive results in identifying and distinguishing between natural, heated, and lab-grown spinel and alexandrite based on luminescence decay profiles [20]. These gemstones demonstrate long photoluminescence lifetimes, with that for spinel ranging from 9 to 23 microseconds and that for alexandrite ranging from 25 to 53 microseconds, which can be quantitatively measured and differentiated (**Figure 10**). Furthermore, studying the fluorescence

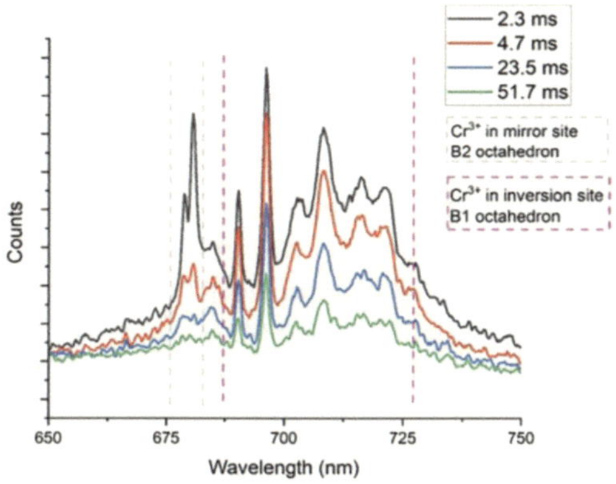

Figure 9.
Time-resolved fluorescence decay spectra of alexandrite, source [20].

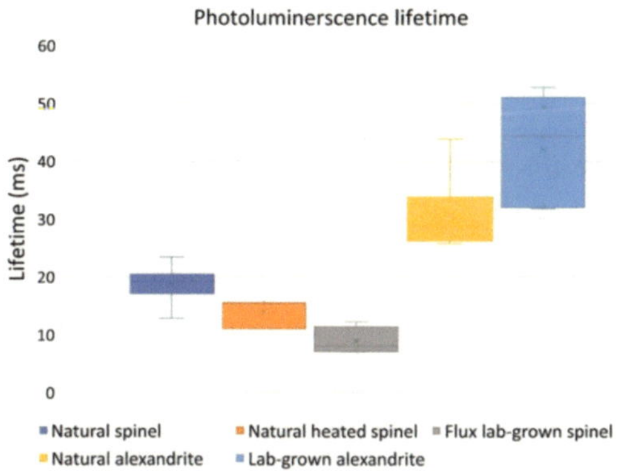

Figure 10.
PL lifetime of spinel and alexandrite varieties, source [20].

decay of ruby, another Cr-doped gemstone, revealed a notably shorter lifetime compared to spinel and alexandrite, suggesting a clear distinction based on crystallographic properties. These findings highlight the importance of structural complexity in governing electron decay paths and the duration of excited states, which can be altered through treatments such as heating, thus affecting the decay profiles.

As an additional non-destructive technique alongside Raman and photoluminescence spectroscopic analyses, photoluminescence lifetime analysis (PLLA) provides deeper insights into the crystal structure. For instance, the study of time-resolved photoluminescence spectroscopy on spinel and alexandrite revealed the Mg-Al order–disorder behavior of spinel post-heating through semi-quantitative fitting analysis of the exponential decay curves. It also successfully laid out the contributions of Cr in alexandrite in two types of slightly distorted interstitial octahedral sites: octahedra with Ci symmetry and octahedra with Cs symmetry [20]. Such intricate studies are typically the domain of those applying more complex analytical techniques, such as X-ray diffraction (XRD) or single-crystal X-ray analysis. From a gemological standpoint, distinguishing heated natural spinel from flux-grown synthetic spinel, or natural alexandrite from synthetic alexandrite, often requires the assistance of advanced trace element analysis techniques such as LA-ICPMS or FTIR. Both Raman and singular photoluminescence spectroscopic analyses reach their limits and are unable to differentiate between them. However, PLLA alters this predicament by offering new avenues from the insight of crystal structure to address these challenges, all while maintaining the non-destructive approach that is paramount in gemological analysis.

LA-ICPMS represents the pinnacle of contemporary gemological testing methodologies, distinguishing itself as an indispensable tool in technologically advanced gemological laboratories. This sophisticated technique primarily focuses on the precise quantitative analysis of trace elements, a critical aspect for determining the provenance of prestigious gemstones such as rubies, sapphires, emeralds, alexandrites, and red spinels. Furthermore, LA-ICPMS excels in identifying complex treatments in gemstones, such as beryllium diffusion in sapphires, and discerning between naturally heated spinels and flux synthetic spinels, as highlighted in this discussed section.

Nonetheless, while LA-ICPMS is renowned for its remarkable precision and comprehensive analytical capabilities, it is important to recognize its limitations. One of the primary challenges lies in its high operational and maintenance costs, alongside its restricted portability, which can be a significant barrier, especially for smaller-scale laboratories. In light of these constraints, the advent of the PLLA method represents a promising shift in the field. PLLA offers a more cost effective and expedient solution, making it an accessible alternative for gemological laboratories that may not have the extensive resources required for LA-ICPMS.

It is crucial to acknowledge, however, that PLLA is still in its nascent stages. The consistency and reliability of results obtained from PLLA as compared to LA-ICPMS necessitate further empirical investigation. This involves the analysis of a larger pool of comparative samples and extensive data to affirm the precision and reliability of PLLA. Investing in such studies is of paramount importance. The primary advantage of Photoluminescence and PLLA lies in their non-destructive nature and rapid analysis capabilities, coupled with significantly lower operational costs. These attributes not only enhance the efficiency of gemological testing but also broaden the scope of their application, making spectroscopical analysis more accessible and sustainable in the long run.

The initial research with the 405 nm PL device offers promising insights into the capabilities of time-resolved photoluminescence in gemstone identification and treatment detection. As we look toward future studies, the use of more refined measurements and sophisticated instrumentation is anticipated to refine the methods for distinguishing between various groups of gemstones.

Unlike traditional testing methods, time-resolved photoluminescence provides quantitative assessments, processing large data sets which, when combined with automated data acquisition and analysis software, can meet the rigorous demands of gemstone testing analysis. The outlook for PLLA in gemstone testing is promising, with potential applications in determining gemstone treatments, identifying species, and possibly tracing origins, thereby advancing both the precision and efficiency of gemological studies.

5. Integration with automation of big data analysis

In this chapter, we have explored laser-induced spectroscopic methods, specifically Raman spectroscopy, photoluminescence analysis, and photoluminescence lifetime analysis (PLLA), and showcase their practical applications in addressing routine challenges faced in gemological laboratories. Many of these techniques and their corresponding equipment are still in the early stages of development, yet they hold the potential to revolutionize the capabilities of non-destructive analytical processes.

A shared characteristic of these methods is the standardized collection of spectral data, which can be automatically processed using sophisticated algorithms. The instruments developed to date, for instance, are equipped with automated spectrum decoding programs [22, 23, 59]. This feature is a significant advancement compared to many traditional gemological testing methods that rely on subjective interpretation. Laser excitation spectroscopy methods allow for objective and independent analysis based on an established database, greatly enhancing the reliability and efficiency of gemological assessments.

As previously discussed, relying on a single method for gemstone testing can lead to deviations in results due to the limitations of the material being tested, the testing method, or the instrumentation itself. In an improvement on that scenario, since Raman spectroscopy, photoluminescence, and PLLA each have their unique strengths, they can complement each other, make up for one another's shortcomings, and cross-verify conclusions. Consider an example with emerald, where a low-quality emerald with excessive fillers causing excessive fluorescence may disrupt the Raman spectrum test. In such instances, the characteristic fluorescence spectrum of chromium in a beryl crystal environment can be used to confirm that it is an emerald as a conclusive result. Conversely, for iron-colored green tourmaline, which may not exhibit any fluorescence spectrum, a clear Raman spectrum of tourmaline can be obtained without interference from fluorescence. In the case of spinel, the overlapping Raman and photoluminescence (PL) spectra between heated natural spinel and flux lab-grown spinel pose a challenge for spectroscopic testing. PLLA becomes crucial in solving this problem. Similarly, the issue of overlapping Raman and PL spectra between natural and lab-grown alexandrite can be effectively addressed using PLLA. When feasible, employing multiple methods simultaneously and compiling an extensive reference database can significantly enhance the accuracy of test results and the determination of mineral subspecies, reducing the need for subjective judgments from gemologists.

The implementation of laser-induced spectroscopic testing systems is notably advantageous due to their compatibility with big data analysis, which centers on a thorough understanding of various data types, the application of flexible data analysis tools, and a focused approach to defining and predicting expected outcomes. These principles are supported by the four critical dimensions of big data, commonly referred to as the four V's [64]: volume, which pertains to the quantity of data; variety, relating to the different types of data; velocity, which addresses the speed of data processing and analysis; and veracity, concerning the accuracy and reliability of the data. This alignment with big data principles enhances the efficacy and precision of these spectroscopic testing systems.

Laser-induced spectroscopic testing systems exemplify these principles in action. These methods are adept at handling the "volume" of data generated from singular or multiple excitation sources, such as a 405 nm laser, as described in the text. This allows for the categorization and sub-categorization of materials, satisfying the "variety" aspect based on a consistent set of analytical conclusions. Meanwhile, the "velocity" of data analysis is unprecedented, with the capability to complete testing across all three methods in a timeframe ranging from a mere 1 to 10 seconds. Finally, the "veracity" of the testing outcomes is ensured through mutual verification across the methods, reinforcing the reliability and accuracy of the conclusions drawn.

This comprehensive approach to data analysis in gemological testing through laser-induced spectroscopy not only streamlines the process but also enhances the precision and trustworthiness of the results. As such, this integration of modern techniques represents a significant leap forward in the field, offering a more nuanced and rapid understanding of gemological properties.

6. Conclusion

This chapter delved into the transformative impact of laser-induced spectroscopy in the realm of gemology. The deployment of Raman spectroscopy, photoluminescence (PL), and photoluminescence lifetime analysis (PLLA) represents a paradigm shift in the way gemstones are analyzed and characterized. To date, these techniques have proven invaluable in distinguishing natural gems from lab-grown counterparts, identifying treatment methods, and tracing the origins of significant colored gemstones.

The integration of these advanced spectroscopic methods in gemological testing has not only enhanced the precision and accuracy of gemstone identification but also streamlined the analysis process. The ability to conduct non-destructive, rapid, and reliable tests on gemstones using these technologies is a significant advancement over traditional methods. This progression aligns seamlessly with the growing need for more sophisticated and efficient testing procedures in the face of the increasing complexity of gemstone treatments and the proliferation of synthetic varieties.

Photoluminescence, in particular, has emerged as a crucial tool for understanding the chemical and structural properties of gemstones. Its application in determining the chromium contents in various stones and its role in identifying heat treatments in minerals such as spinel and sapphire exemplify its versatility and effectiveness. Similarly, Raman spectroscopy's capacity to classify minerals accurately and its application in distinguishing natural from synthetic gemstones underscore its significance in modern gemology. Meanwhile, photoluminescence lifetime analysis (PLLA) provides deeper insights into the crystal structure of gemstones. This method opens up a new avenue for investigating intricate details of alterations in the crystal lattice in a non-destructive manner.

Furthermore, the adoption of these techniques in gemological laboratories has facilitated greater transparency and accountability within the industry. In this sense, by providing more detailed and accurate information about gemstones, these methods assist in maintaining the integrity of the market and the trust of consumers.

As we look to the future, the ongoing development and integration of these advanced spectroscopic techniques with automation and big data analysis promise to further revolutionize gemological testing. This evolution will likely lead to even more sophisticated, efficient, and comprehensive analysis methods, ensuring that the gemology field remains at the forefront of scientific innovation and continues to meet the challenges posed by an ever-changing market.

The advancements discussed in this chapter not only highlight the current state of gemological testing but also pave the way for future innovations that will continue to enhance the field's ability to discern, categorize, and appreciate the intricate beauty and complexity of gemstones.

Acknowledgements

I would like to extend my heartfelt thanks to my former colleagues at the GIA's research department: Dr. T.H. Tsai, Dr. A. Palke, and Dr. W. Wang. Their willingness to involve me in groundbreaking research and innovation projects, coupled with enriching professional knowledge exchanges.

Conflict of interest

The authors declare no conflict of interest.

Author details

Wenxing Xu
Independent Researcher, Lucerne, Switzerland

*Address all correspondence to: wenxingx@gmail.com

References

[1] Willems B, Tallaire A, Barjon J. Exploring the origin and nature of luminescent regions in CVD synthetic diamond. Gems & Gemology. 2011;**47**(3):202-207

[2] Eaton-Magaña S, Shigley JE. Observations on CVD-grown. Gems & Gemology. 2016;**52**(3):222-245

[3] Eaton-Magaña S, Shigley JE, Breeding CM. Observations on HPHT-grown synthetic diamonds: A review. Gems & Gemology. 2017;**53**:262-284

[4] Lu Q, Gong H, Guo Q, Huang X, Cai J. Gemological characteristic difference between colorless CVD synthetic diamonds and natural diamonds. Materials. 2021;**14**(20):6225. DOI: 10.3390/ma14206225

[5] Moses TM, Reinitz I, Fritsch E, Shigley JE. Two treated-color synthetic red diamonds seen in the trade. Gems & Gemology. 1993;**29**:182-190

[6] Fisher D, Spits RA. Spectroscopic evidence of GE POL HPHT-treated natural type IIA diamonds. Gems & Gemology. 2000;**36**:42-49

[7] Collins AT. The colour of diamond and how it may be changed. The Journal of Gemmology. 2001;**27**:341-359

[8] Martineau PM, Lawson SC, Taylor AJ, Quinn SJ, Evans DJF, Crowder MJ. Identification of synthetic diamond grown using chemical vapor deposition (CVD). Gems & Gemology. 2004;**40**:2-25

[9] Shigley JE, McClure SF, Breeding CM, Shen AH-T, Muhlmeister SM. Lab-grown colored diamonds from Chatham created gems. Gems & Gemology. 2004;**40**:128-145

[10] Shigley JE, Breeding CM, Shen AH-T. An updated chart on the characteristics of HPHT-grown synthetic diamonds. Gems & Gemology. 2004;**40**:303-313

[11] Wang W, Doering P, Tower J, Lu R, Eaton-Magaña S, Johnson P, et al. Strongly colored pink CVD lab-grown diamonds. Gems & Gemology. 2010;**46**:4-17

[12] Wang W, D'Haenens-Johansson UFS, Johnson P, Soe Moe K, Emerson E, Newton ME, et al. CVD synthetic diamonds from Gemesis Corp. Gems & Gemology. 2012;**48**:80-97

[13] Wanthanachaisaeng B, Bunnag N, Sutthirat C, Atichat W, Ounorn P, Sripoonjan T. Luminescence of beryllium heat treated corundum. In: Burapha University International Conference. Thailand: Burapha University; 2012

[14] Wanthanachaisaeng B, Bunnag N, Sutthirat C, Atichat W, Ounorn P, Sripoonjan T, et al. Investigation of Be-treated sapphire by luminescence spectroscopy. In: in Conference on Raman and Luminescence Spectroscopy CORALS–2013. Austria: University of Vienna; 2013

[15] Eaton-Magaña S, Breeding CM. An introduction to photoluminescence spectroscopy for diamond and its applications in gemology. Gems & Gemology. 2016;**52**:2-17

[16] IGR. Basic Elements of Photoluminescence Spectroscopy in Gemology [Online]. IGR; 2018. Available from: https://www.rivistaitalianadigemmologia.com/

en/2018/02/23/basic-elements-of-photoluminescence-spectroscopy-in-gemology/

[17] Eaton-Magaña S, Breeding CM, Palke AC, Homkrajae A, Sun A, McElhenny G. Raman and photoluminescence mapping of gem materials. Minerals. 2021;**1**(2):177. DOI: 10.3390/min11020177

[18] Gupta A, Hacquebard L, Childress L. "Efficient signal processing for time-resolved fluorescence detection of nitrogen-vacancy spins in diamond". Journal of the Optical Society. 2016;33: 28-34

[19] Jones D, Alexandrov Y, Curry N, Kumar S, Lanigan P, McGuiness C, et al. Multidimensional spectroscopy and imaging of defects in synthetic diamond: Excitation-emission-lifetime luminescence measurements with multiexponential fitting and phasor analysis. Journal of Physics D: Applied Physics. 2021:5303

[20] Xu W, Tsai T-H, Palke A. Study of 405 nm laser-induced time-resolved photoluminescence spectroscopy on spinel and alexandrite. Minerals. 2023;**13**(3):419

[21] Scarani A, Åström M. Raman spectroscopy: Technique. Rivista Italiana Di Gemmologia. 2017;**2**:42-45

[22] Tsai T, Xu W. Rapid gemstone mineral identification using portable Raman spectroscopy. Journal of Raman Spectroscopy. 2023;**54**:640-650

[23] Tsai T-H, D'Haenens-Johansson UFS. Rapid gemstone screening and identification using fluorescence spectroscopy. Applied Optics. 2021;**60**:3412-3421

[24] GIA. Gem Identification Lab Manual. Carlsbad, California: Gemological Institute of America; 2005

[25] Groat L, Giuliani G, Stone-Sundberg J, Renfro ND, Sun Z. A review of analytical methods used in geographic origin determination of gemstones. Gem & Gemology. 2019;**55**:512-535

[26] Liddicoat R. Developing the powers of observation. Gems & Gemology. 1962;**10**:291

[27] Fritsch E, Rossman GR. An update on color in gems. Part 1: Introduction and colors caused by dispersed metal ions. Gems & Gemology. 1987;**23**:126-139

[28] Chow BHY, Reyes-Aldasoro CC. Automatic gemstone classification using computer vision. Minerals. 2022;**12**(1):60

[29] Breeding C. Developments in gemstone analysis techniques and instrumentation during the 2000s. Gems & Gemology. 2010;**24**:241

[30] Anton-paar. Handheld Raman Spectrometer: Cora 100 [Online]. Austria: Anton-paar; 2024. Available from: https://www.anton-paar.com/ca-en/products/details/handheld-raman-spectrometer-cora-100/

[31] StellarNet, Inc. StellarRAM Handheld Raman Spectrometer [Online]. Florida: StellarNet, Inc.; 2024. Available from: https://www.shopstellarnet.com/stellarram-handheld-raman-spectrometer/

[32] StellarNet Inc. 785nm Preconfigured Raman Spectrometer System [Online]. Florida: StellarNet Inc.; 2024. Available from: https://www.shopstellarnet.com/785nm-preconfigured-raman-spectrometer-system/

[33] Magilabs Ltd. gemmoraman. Magilabs Ltd.; 2024 [Online]. Available from: https://www.gemmoraman.com/products/gemmoraman-532/

[34] Krishnan R. Raman Spectrum of diamond. Nature. 1945;**155**:216

[35] Beny JM, Piriou B. Vibrational spectra of single-crystal topaz. Physics and Chemistry of Minerals. 1987;**15**:148-159

[36] Saeseaw S, Wang W, Scarratt K, Emmett JL, Douthit TR. Distinguishing Heated Spinels from Unheated Natural Spinels and from Synthetic Spinels. 2009. [Online]. Available from: https://www.gia.edu/doc/distinguishing-heated-spinels-from-unheated-natural-spinels.pdf

[37] Huong LT-T, Häger T, Hofmeister W. Confocal micro-Raman spectroscopy: A powerful tool to identify natural and synthetic emeralds. Gems & Gemology. 2010;**46**:36-41

[38] Illy E, Karlsson H. How to Choose a Laser: How to Choose a Laser for Raman Spectroscopy. Oklahoma: Laser Focus World; 2018. [Online]. Available from: https://www.laserfocusworld.com/lasers-sources/article/16555207/how-to-choose-a-laser-how-to-choose-a-laser-for-raman-spectroscopy

[39] Metrohm. Auswahl der am besten geeigneten Laser-Wellenlänge für Ihre Raman-Anwendung. 2024. [Online]. Available from: https://www.metrohm.com/de_ch/applications/bw-tek-applikationen/410000001-c.html

[40] CNI. 405 nm Violet Blue Diode Laser. Changchun, China: CNI; 2024 [Online]. Available from: http://www.cnilaser.com/blue_laser405.htm

[41] Renishaw. Multi-laser Raman analysis of zirconia. 2017. [Online]. Available from: https://www.renishaw.com/en/multi-laser-raman-analysis-of-zirconia--43004

[42] UNSW, "Renishaw. inVia Reflex. Raman Microscope (532, 785, 830 and 1064 nm)," 2024. [Online]. Available from: https://www.analytical.unsw.edu.au/facilities/speclab/instruments/renishaw-invia-reflex-raman-microscope-532-785-830-and-1064-nm

[43] Stockton CM. The separation of natural from synthetic emeralds by infrared spectroscopy. Gems & Gemology. 1987;**23**:96-99

[44] Mashkovtsev RI, Solntsev VP. Channel constituents in synthetic beryl: Ammonium. Physics and Chemistry of Minerals. 2002;**29**(1):65-71

[45] Schmetzer K, Kiefert L. Water in beryl—A contribution to the separability of natural and synthetic emeralds by infrared spectroscopy. Journal of Gemmology. 1990;**22**:215-223

[46] Renishaw. Photoluminescence spectroscopy and fluorescence explained. renishaw, [Online]. Available from: https://www.renishaw.com/en/photoluminescence-spectroscopy-and-fluorescence-explained--25809

[47] Tuschel D. Photoluminescence spectroscopy using a Raman spectrometer. Spectroscopy. 2016;**31**:14-21

[48] Tuschel D. Selecting an excitation wavelength for Raman spectroscopy. Spectroscopy. 2016;**31**:14-23

[49] Gaft M, Reisfeld R, Panczer G. Modern Luminescence Spectroscopy of Minerals and Materials. Cham, Switzerland: Springer; 2015

[50] Sicree A. Chrome, Rubies, Emeralds, and Alexandrite. 2007. [Online]. Available from: http://worcestermineralclub.org/wp-content/uploads/2015/03/PopMin-07a.pdf

[51] Gemporia. Gemstone Elements: Chromium. United Kingdom:

Gemporia; 2017 [Online]. Available from: https://www.gemporia.com/en-gb/gemology-hub/article/644/gemstone-elements-chromium/

[52] Peretti A, Günther D, Haris M. New type of treatment of spinel discovered involving heat-treatment and cobalt-diffusion. 2015. [Online]. Available from: https://www.gemresearch.ch/assets/documents/publication-articles/2015-05-spinel-diffusion-treatment-1.pdf

[53] Yamanaka T, Takéuchi Y. Order-disorder transition in $MgAl_2O_4$ spinel at high temperatures up to 1700°C. Zeitschrift fuer Kristallographie. 1983;**165**:65-78

[54] Peterson R, Lager G, Hitterman R. A time-of-flight powder diffraction study of $MgAl_2O_4$ at temperatures up to 1273 k. American Mineralogist. 1991;**76**:1455-1458

[55] Redfern S, Harrison R, O'Neill H, Wood D. Thermodynamics and kinetics of cation ordering in $MgAl_2O_4$ spinel up to 1600°C from *in situ* neutron diffraction. American Mineralogist. 1999;**84**:299-310

[56] Andreozzi G, Princivalle F. Kinetics of cation ordering in synthetic $MgAl_2O_4$ spinel. American Mineralogist. 2002;**87**:838-844

[57] Méducin F, Redfern S. Study of cation order-disorder in spinel by in situ neutron diffraction up to 1600 K and 3.2 GPa. American Mineralogist. 2004;**89**:981-986

[58] Princivalle F, Martignago F, Dal Negro A. Kinetics of cation ordering in natural $Mg(Al, Cr^{3+})2O_4$ spinels. American Mineralogist. 2006;**91**:313-318

[59] Tsai T-H. Multi-excitation fluorescence imaging for identifying clarity enhancement in gemstones. In: Proc. SPIE 11815, Novel Optical Systems, Methods, and Applications XXIV. Vol. 1181505. Washington, USA: SPIE (The International Society for Optics and Photonics); 2021

[60] Arem JE. What Can Gemstone Luminescence Indicate? IGS; 2024. [Online]. Available from: https://www.gemsociety.org/article/understanding-luminescence-gemology/

[61] Palke AC, Saeseaw S, Renfro ND, Sun Z, McClure SF. Geographic origin determination of ruby. Gems & Gemology. 2019;**55**:580-612

[62] Palke AC, Saeseaw S, Renfro ND, Sun Z, McClure SF. Geographic origin determination of blue sapphire. Gems & Gemology. 2019;**55**:536-579

[63] Lanigan PMP, McGuinness CD, Rendle M, Aked PA, Bearcroft CG, Jones DC, et al. Real-time detection of long lived near infrared luminescence from colourless cubic zirconia by time-gated imaging. Minerals. 2020;**10**:891-902

[64] Opensistemas. The Four V's of Big Data. 2023. [Online]. Available from: https://opensistemas.com/en/the-four-vs-of-big-data/

Chapter 7

Recent Advances in Machine Learning Methodologies for LIBS Quantitative Analysis

Hao Liu, Kai Han, Weiqiang Yang and Minsun Chen

Abstract

The mapping between LIBS spectral data to the quantitative results can become highly complicated and nonlinear due to experimental conditions, sample surface state, matrix effect, self-absorption, etc. Therefore, the accurate quantitative analysis is the longstanding dream of the LIBS community. The advantages of machine learning in dealing with high-dimensional and nonlinear problems have made it a cutting-edge hot topic in quantitative LIBS in recent years. This chapter introduces the current bottlenecks in quantitative LIBS, sorts out the data processing methods, and reviews the research status and progress of conventional machine learning methods such as PLS, SVM, LSSVM, Lasso, and artificial neural network-based methods. By comparing the results of different methods, the perspective of future developments on learning-based methods is discussed. This chapter aims to review the applications of the combination of quantitative LIBS and machine learning methods and demonstrate the performance of different machine learning methods based on experimental results.

Keywords: laser-induced breakdown spectroscopy, machine learning, quantitative analysis, artificial neural network, LIBS data processing, AI for science

1. Introduction

Laser-induced breakdown spectroscopy (LIBS) is a noninvasive spectroscopy technique based on the emission from the excitation of atoms in materials, and it is widely applied for elemental compositional analysis. The sketch of the LIBS is illustrated in **Figure 1**. The spectra and images of the plasma plume produced by the pulsed laser ablation onto the sample surface are collected by the spectral-resolved measurement devices. The electrons and ions/atoms ejected from the sample can represent the quantitative characteristics (e.g. elements, concentrations, etc.) of the target, and LIBS offers a spectrometric method to obtain the information of samples.

The quantitative results can be obtained by analyzing the spectrum emitted from the plasma produced by a pulsed laser irradiation. Due to its capabilities of standoff and online measurements, portability, quasi nondestructive detection, etc., LIBS has widely used in the fields of environmental monitoring, space missions, industrial

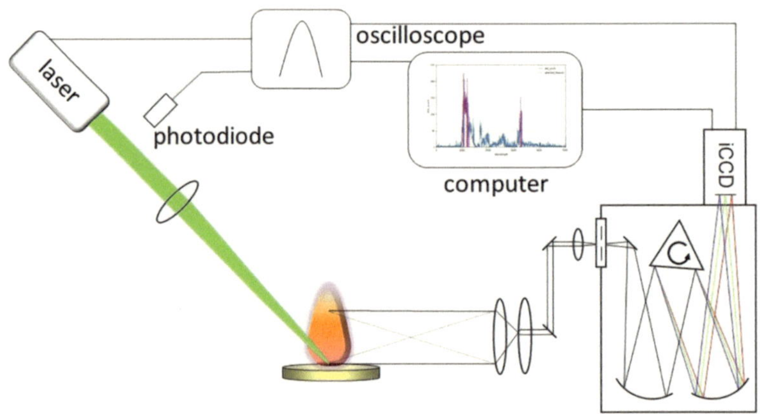

Figure 1.
The sketch of the LIBS method.

monitoring, and isotope analysis. Several excellent existing works have reviewed LIBS applications [1–4], mainly focusing on specific areas or techniques such as diagnosis of the plasma-facing wall of confinement fusion devices [5–7], industrial applications [8–10] and data analysis [11], nanoparticles in LIBS [12, 13], underwater applications [14], aerosol analysis [15], and the study of uranium-containing compounds [16].

Since the physics and chemical mechanism involving complicated processes, including laser-sample target, laser-plasma, and plasma-sample interactions, the emission spectra are sensitive to laser parameters, ambient gas pressure, target characteristics (e.g. heat capacity, thermal conductivity, surface roughness, wavelength-dependent absorptivity, etc.). Their temporal and spatial evolving [17, 18] lead to complex nonlinear mapping from spectral data to quantitative results, thus, the accuracy and real-time performance of quantitative LIBS needs further improvements. In recent years, the rapid developments of machine learning and deep learning methods have received extensive attention from LIBS community due to their brilliant capability for solving nonlinear problems. Chen et al. [19] summarized machine learning methods in the LIBS analysis of geological applications. Li et al. [3] reviewed artificial neural networks (ANN) methods in geology, biology, and industrial applications, demonstrating the potential of ANN-based LIBS for material identification/classification and quantitative analysis. Zhang et al. [11] focused on evaluating the performance of conventional machine learning methods, such as principal component analysis (PCA) and support vector machines (SVM), on LIBS classification and regression. Huang et al. [20] provided a review of machine learning-based LIBS in soil analysis tasks.

These works concluded the comprehensive application of machine learning methods in LIBS classification and regression tasks; however, quantitative analysis, especially accurate quantitation, is still the vital barrier for LIBS toward real-world practicality. Since different types of data processing methods and machine learning methods have their own strength and inadequacy, it is of great importance to analyze and optimize one or hybrid appropriate methods according to application scenes and tasks. While at present, few reviews specifically focus on the comparative analysis and outlook of learning-based methods for quantitative LIBS. Following prior excellent reviews, this chapter emphasizes on machine learning techniques in quantitative analysis and summarizes the existing methods and advances in combination with the latest artificial intelligence (AI) results and development trends. Based on the latest experimental results, we propose a perspective on the future development of learning-based quantitative LIBS from the viewpoint of AI-driven LIBS research.

2. LIBS data preprocessing methods

LIBS system usually includes pulsed lasers, laser focusing optics, spectral collection optics, spectrometers, spectral calculation, and interpretation devices. Nanosecond lasers are the most common-utilized LIBS light sources in laboratory and industrial scenarios due to its advantages of small size and weight, low cost, and portability. In recent decades, with the development of ultrafast laser technology, femtosecond lasers have also been applied to LIBS analysis with benefit of their characteristics, such as smaller ablation volume and cold ablation effect [21].

Although the LIBS hardware system seems to be relatively simple at first glance, the laser characteristics (e.g., wavelength, pulse width, energy), focusing characteristics (e.g., spot size, ablation rate, incident angle), target characteristics (surface roughness, absorption rate, thermal conductivity, etc.), and ambient environment characteristics (e.g., background gases and their pressures) during the process [17, 22–26] significantly affect the plasma emission spectra, reducing the accuracy elements concentration calculations by analyzing spectra emitted from laser-induced plasma. Despite its strength of online measurements, LIBS with conventional quantitative methods has a lower limit of detection (LOD) in comparison to those offline chemometric techniques. For trace element detection in most solids, the LIBS LOD is often in the range from 1 to 100 ppm [27], which barely meets the needs of several element trace tasks.

In order to enhance the reproducibility of LIBS spectral signals and improve the signal-to-noise ratio, spatial confinement [28], electric-field-assisted [29], magnetic-field-assisted [30], polarization enhancement [31], nanoparticle enhancement [32], and optical capture enhancement [12] have been proposed in the current study to establish stable experimental acquisition conditions in order to improve the quality of raw data. For the acquired spectral data, data preprocessing is one of the key steps for quantitative analysis of LIBS, which can effectively improve the accuracy and reproducibility of the results. Existing spectral data preprocessing methods can be classified into four categories: accumulation and averaging, data filtering, normalization and standardization, and self-absorption calibration.

2.1 Accumulation and averaging

Accumulating and averaging measured multiple spectra is effective in reducing noise and improving repeatability and signal-to-noise ratio (SNR). Theoretically, accumulating multiple spectra can improve the spectral SNR while avoiding spectral intensity saturation, and the higher the number of accumulations, the more beneficial it is to the quantitative results. However, according to the principle of statistics, if the noise affecting spectral repeatability obeys a normally distributed random noise with expectation m and standard deviation s, the expectation of the average intensity and standard deviation of the noise of the accumulated n spectral measurements are also m and s. Therefore, the larger the number of accumulations is, the smaller impact of further increase in the number of accumulations on the improvement of signal repeatability, that is. the marginal utility is decreasing.

2.2 Data filtering

Data filtering methods mainly include smoothing, mean filtering, Kalman filtering, Wiener filtering, adaptive filtering, and wavelet transform. These methods can theoretically divide the spectra into "*signal*" and "*noise,*" and then remove the "*noise*" part from the spectra to improve the SNR and enhance the reproducibility of the LIBS plasma and its spectra, hence the reproducibility of the LIBS plasma and its spectra is enhanced. However, it is very difficult to define "*signal*" and "*noise*" in real scenarios. For example, in steel samples, the characteristic spectral signals of trace elements, such as sulfur and phosphorus, which are usually in the order of 0.01% or less, can be easily ignored and regarded as the "*noise*" signal due to the insufficient SNR. Therefore, before applying filtering methods to spectral data, it is important to consider their physical background and their impact on the raw LIBS data. It is often possible to optimize the filtering method and determine the filtering parameters based on the physical mechanism of quantitative analysis. For example, moving averages and the resulting moving convolutions that can be adapted to machine learning methods are more typical spectral smoothing methods. The method can be applied to the spectra of each sample, thus reducing errors due to instrumental and environmental noise. It is important to note that the choice of the sliding window width is one of the most important parameters. If the moving window is too narrow, the smoothing effect will be limited, while if the moving window is too wide, the resolution of the spectrum will be compromised, resulting in a poor quantitative analysis.

2.3 Normalization and standardization

Spectral normalization, also known as spectral standardization, is a commonly used method to reduce signal uncertainty and improve signal concentration correlation. Normalization methods include background normalization, spectral integral normalization, standard normal variable normalization, internal standard normalization, and external standard normalization [33]. Usually, normalizing the background can effectively improve the data quality, but some of the findings also give contrast results [34].

Spectral integral method divides the spectrum intensity by the integral intensity of the whole spectrum to obtain the normalized spectrum [35]. Standard normal variational method refers to the intensity calibration to the standard normal distribution [36]. Internal standard method involves selecting the reference spectral lines in the vicinity of the interested spectral lines to calculate the relative intensities [37], and external normalization uses spectral lines from the known elemental concentration to calibrate the spectral lines of unknown elements.

Several studies on LIBS normalization are listed in **Table 1**. Some prior research hypothesized that normalization is expected to yield better experimental results but did not verify it by comparing experiments.

From the table, the R^2 is often used as an evaluation parameter, complemented by other metric parameters such as root mean square error (RMSE), root mean square error of the validation set (RMSEC), root mean square error of the cross-validation (RMSECV), and the limit of detection (LOD), which allow evaluation of the superiority from different perspective.

In addition, some studies have used other parameters from LIBS experiments as normalization references. For example, a normalization method based on plasma parameters is proposed and applied to archeological samples [42]. The total number

Normalization method	Norm vs. non-norm comparison	Metric parameters	Reference
internal normalization	Yes	RMSECV, LOD	Thomas et al. [38]
internal normalization	Yes	RMSEC, R^2	Andrade et al. [39]
background normalization	No	R^2, LOD	Dell'Aglio et al. [34]
spectral integral	No	RMSE, R^2	Payre et al. [40]
spectral integral	Yes	RMSECV, R^2	Takahashi et al. [41]
standard normal	Yes	R^2, LOD	Syvilay et al. [36]

Table 1.
Several existing normalization methods, where RMSECV is root mean square error of cross-validation, RMSEC is root mean square error of cross-validation, RMSE is root mean square error, and LOD is the limit of detection.

density, temperature, and electron number density can also be used to normalize the measured spectra under different conditions in order to compensate for fluctuations in the spectral signals due to variations in the nature of the plasma [43]. A method based on the ideal sample normalization is proposed by [44]. The method assumes that the plasma temperature (T), electron number density (N_e), and atomic number density (N_s) of the measured element are at standard values in the ideal state, and each actual measured signal can be regarded as the ideal state value plus the deviation caused by the variation of T, N_e, and N_s. By converting each measured spectrum to a spectrum at the standard state, the measurement uncertainty can be reduced. Similarly, researchers have used the current of the LIBS plasma, total emission intensity, plasma images, the Euclidean norm, the maximum and minimum intensity values of each individual spectrum [45–47], and other reference signals to calibrate spectral raw data [48].

However, for quantitative analysis, the plasma does not always satisfy the local thermodynamic equilibrium (LTE). In addition, the elements with low concentration in the target yield a small intensity of the spectral lines, so the SNR is not sufficient to derive accurate quantitative results. Thus, the use of calibration methods in practical applications needs careful assessment based on the actual state of the LIBS plasma. In conclusion, although normalization can reduce the fluctuation and error of the signal under certain circumstances, the normalization method that lacks mechanism support can also lead to a worse quantitative accuracy than the non-normalized case [34]. Therefore, the selection of normalization methods should consider availability of prior knowledge, suitable reference lines, and reliable mechanistic models.

2.4 self-absorption calibration

The spectrum of spontaneous radiation in the central region of the plasma undergoes self-absorption as it passes through the region occupied by the plasma. In the measurement of relative concentration of trace elements to iron in steel samples by using the remote LIBS technique [49], nonlinear effects are still obvious due to the self-absorption even after the internal normalization. Experiments [50] have shown that even if no significant attenuation in the line shape, the self-absorption effect still presents and affects the accuracy of the quantitative results. Theoretical models for optically thick plasmas are commonly used in studies to calibrate self-absorption *via* the curve of growth (COG). The COG method is firstly applied to the Cr peak and analyzed the properties of the laser-induced plasma, such as temperature. The mode [51] l demonstrated the difference in the relationship between peak intensity and concentration over a wide range of concentrations. The line absorption model [52] is also used to calculate theoretical intensities from the significant self-absorbing Cu lines, demonstrating that their method is effective for generating highly linear calibration curves of the Cu/Si intensity ratio as a function of Cu concentration. Lazic et al. [53] calculated optically thick plasma self-absorption of different elemental spectral lines and showed that self-absorption can lead to relative errors of up to 20% in quantitative LIBS analysis.

The temporal–spatial evolution of LIBS plasma is complicated [25], so the self-absorption study is important to understand the mechanism of plasma emission spectra in LIBS. Existing studies have proposed many methods to reduce matrix effects and correct for self-absorption to improve the accuracy of calibration quantitative analysis. However, the effectiveness of these methods still depends on the samples and experimental conditions used. For example, for the quantitative analysis of samples with complex compositions, such as steel, rock, and soil, the nonlinear effects caused by matrix effects and self-absorption effects are still the bottleneck, which limits the high-precision quantitative analysis of LIBS.

3. Conventional machine learning methods for LIBS

3.1 Partial least squares (PLS)

Partial least squares (PLS) regression is a statistical-based machine learning method that obtains a linear regression model by projecting the prediction and observable variables onto a new space and then determining maximum variance hyperplane between the function and the independent variables. PLS has been widely used in the quantitative LIBS, especially in the quantitative analysis of samples containing multiple elements, for example, the prediction of chemical composition of rock samples [54]. The serial partial least squares (S-PLS) and multiblock partial least squares (MB-PLS) [55] are used in quantitative analysis. The hybrid method [56] of PLS and wavelet transform is proposed to measure the C content in 24 bituminous coal samples. Since the output of the PLS method is not limited to elemental concentrations, it has also been applied to other quantitative analyses. For example, C element, ash, and volatile components in 58 coal samples are determined by a combined model [57] of principal component analysis and partial least squares. The isotopic ratio of U235/U238 is outputted with relative error of 0.1–8% [58], whose accuracy can meet the demand of in situ detection for industrial production.

The PLS method is not only used for the detection of samples in atmospheric environment but is also applied to underwater LIBS measurements. Large fluctuations occurred in the signal of LIBS in water by PLS and improved the accuracy of the quantitative analysis, which resulted in a reduction of the RMSECV by 30% [59]. In addition, the PLS method has been applied to the *in situ* analysis of Martian rocks by ChemCam, a LIBS device on board Curiosity [60, 61]. Experimental results in laboratory and practical environments [62–65] show that PLS quantitative analysis achieves better results with a 9% reduction in RMSEP compared to other multiple regression methods, such as elastic net, least absolute contraction, and selection operators, support vector regression and k-nearest neighbor regression [66, 67].

Since PLS has theoretical advantages in dealing with multivariate regression problems, the PLS method outputs more accurate results under the circumstance that more variables than observed condition numbers in the prediction matrix, or when there is multicollinearity among the data. It is important to note that PLS relies on statistical correlation or curve fitting, ignoring the physical knowledge of the LIBS measurements, which may lead to excessive noise and ultimately undermine the accuracy of quantitative measurements if the samples are out of the range of calibration sample set. Therefore, PLS should be used in combination with methods that incorporate physical mechanisms, such as principal component analysis and wavelet transform, to obtain better analytical results.

3.2 Support vector machine (SVM)

SVM is a machine learning method based on the principle of structure minimization, which is suitable for multivariate regression problems and has strong adaptability to high-dimensional, complex, and nonlinear data. SVM is widely used in quantitative LIBS, and fusion of principal component analysis (PCA), which reduces the dimensionality of the data, with SVM is a common approach in LIBS.

The combination of PCA and SVM can analyze ash, volatile matter content, and calorific value in 550 kinds of coal samples [68] and can establish calibration model for 35 kinds of coal samples after spectral normalization by Lorentzian and linear functions [69]. Experiments [70] reveal that SVM method could effectively reduce the model complexity and obtain more accurate results of silicon, magnesium, calcium, iron, and aluminum elemental concentrations. However, SVM is not robust to noise, it needs to consume a lot of computational resources and the speed is slow when regression analysis is performed on spectral data embedded with different kinds of noise in the real scene. A combination [71] of particle swarm optimization (PSO) algorithm with the SVM algorithm is proposed to measure the concentration of heavy metal chromium Cr in pork, and both R^2 and RMSE are reduced compared to the traditional SVM method. Compared with the artificial neural network-based quantitative analysis presented in the following section, the SVM method has the advantages of simple parameter control and high interpretability.

Least squares support vector machine (LSSVM) is an improvement of SVM. In LSSVM, the least squares method is used to solve the problem of classification or regression instead of the convex optimization problem in SVM. Thus, LSSVM has a simpler solution process, faster computation and can handle high-dimensional sparse data. In addition, LSSVM can optimize the performance of the model by adjusting the regularization parameters and the parameters of the kernel function. Results of LIBS quantitative analysis on Al-Cu-Mg-Fe-Ni alloy samples [72] show better performance of LSSVM method than PLS regression from the perspective of prediction accuracy, model robustness, computational time, and generalization ability. Prior work [73] also quantitatively analyzed atmospheric deposition of four metallic elements (Pb, Cu, Zn, and Al) by fusion of random forest with the LSSVM method. The predictive performance of different models was evaluated *via* metric parameters such as accuracy, sensitivity, precision, and specificity, and the LSSVM model showed superiority in pollution sources LIBS experiments.

3.3 Least absolute shrinkage and selection operator (LASSO)

In statistics and machine learning, the least absolute shrinkage and selection operator (LASSO) is an often-used regularization-based regression analysis method. Its basic principle is to constrain the complexity of a linear model by applying an L1 regularization term to the model to improve the predictive accuracy and interpretability of the generated statistical model. The Lasso regression can be generally expressed as

$$\hat{\beta} = \underset{\beta}{\operatorname{argmin}} \left\{ \sum_{i=1}^{N} \left(y_i - \beta_0 - \sum_{j=1}^{P} x_{ij} \beta_j \right)^2 + \alpha \sum_{j=1}^{P} \left\| \beta_j \right\|_1 \right\} \tag{1}$$

The advantage of using these linear methods is to prevent overfitting due to the application of penalty terms. The method can be used even if the number of predictions in the dataset is greater than the number of observations. It is also computationally efficient in terms of learning and prediction. The Lasso algorithm can be used to optimize the feature parameter selection in machine learning by ranking the correlation between the different elemental concentration in the sample and the spectral intensity. The larger the absolute value of the Lasso coefficient β indicates the higher correlation between the intensity and the concentration of the element. Therefore, it is very suitable for quantitative LIBS for those samples with many elemental species and large differences in concentration. For example, Boucher et al. [67] used the LASSO regression method to quantitatively analyze the principal components of rocks. Zhang et al. [74] quantitatively analyzed the C element in steel samples, and the relative error was reduced to 13.6%.

4. Artificial neural network (ANN)

Neural network is a powerful and effective analytical method for describing the mapping relationship between the input parameters I, $I \in \mathbb{I}^m$ (m-dimensional feature space $\mathbb{I}$) and the output result O, $O \in \mathbb{O}^n$ (n-dimensional result space $\mathbb{O}$). The method can model complicated systems, including causal and logical reasoning based on knowledge of physics, chemistry, etc., and stochastic processes such as noise by using datasets to train artificial neural network (ANN) and finally accomplish regression analysis task. Specifically for LIBS quantitative analysis, neural networks can construct models to learn the mapping relationship between input parameters such as spectral intensity, spectral line shape, spectral center wavelength, plasma characteristic parameters, experimental condition parameters and output parameters such as the concentration of each element in the sample, containing spectroscopic processes based on physicochemical mechanisms, matrix effects, self-absorption, background noise, random noise, signal fluctuations, and any processes that interfere with the linear/nonlinear relationship between input–output parameters. With sufficient data and appropriate training, neural networks can theoretically achieve high-precision quantitative analysis for complex LIBS processes (**Figure 2**).

4.1 Back-propagation neural network (BPNN)

BPNN is a kind of ANN based on error back-propagation algorithm. BPNN can adaptively learn complex nonlinear relationships, the training speed is fast, and the effect is more prominent when the amount of data is not large. As a supervised learning method, BPNN needs some known samples for training and then uses some unknown samples to test the performance of the training network. For the quantitative analysis of elemental concentration, all samples are usually divided into training,

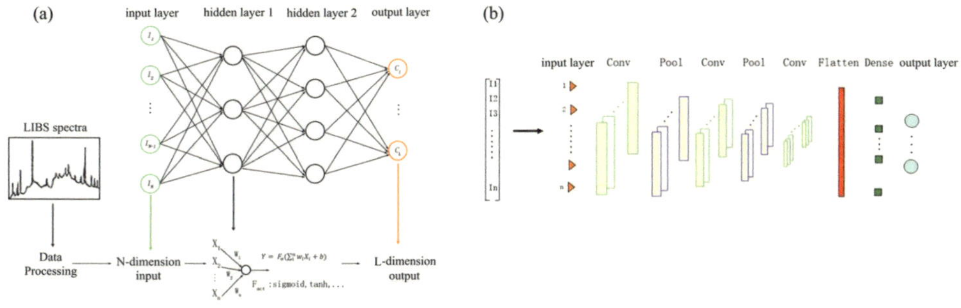

Figure 2.
The common-used ANN structures. (a) Sketch of BPNN, (b) sketch of CNN.

validation, and test samples, and accordingly, the whole LIBS spectral dataset is divided into training, validation, and test sets. The initial weights and bias values of the original BPNN are randomly assigned. When BPNN training is performed, the spectra of the training samples are input into the network along with their concentration truth values. The output concentration values predicted by the BPNN are then compared with the measured values and the total error between them is calculated, and this is iterated for many rounds until the total error converges to the desired value or the number of iterations reaches the pre-set upper limit, thus obtaining a set of final weights and bias values that constitute the convergence of the trained network to the appropriate range. Finally, the predictive ability of the evaluated model is examined by validation and test sets. The trained network predicts and calculates the elemental content in the sample, and its accuracy can be assessed by the difference between the actual and predicted values.

As mentioned above, the problem of the reproducibility of LIBS raw data spectra and the fluctuation of experimental parameters, such as laser energy and plasma temperature, can seriously affect the accuracy of the calculation of the elemental content of the samples. Researchers have solved the problem by building BPNN networks containing experimental parameters. For example, Wang et al. [75] proposed a method for quantitative analysis of LIBS based on key parameter monitoring and back-propagation neural network, called KPBP [76], which used the BPNN algorithm to fit the spectral intensities, and standardized the spectral segments containing the characteristic lines using KPBP. In the study, KPBP experiments were first carried out on the spectra of monolithic samples such as pure aluminum, monocrystalline silicon, and pure zinc in order to optimize the KPBP model. The results show a significant reduction in the RSD of the quantitative results obtained by KPBP compared to the conventional machine learning method and the BPNN method.

4.2 Multilayer perceptron (MLP)

MLP is a common feedforward neural network. It consists of multiple neurons connected through multiple layers. Its basic principle is to pass the output obtained by weighted summing the inputs of each node to the nodes in the next layer through the activation function, and at the same time, to perform nonlinear transformations and dimensionality reduction on the data, to better characterize the structure of the data, and to deal with the complex nonlinear problems. Compared with BPNN in quantitative LIBS, MLP has higher model complexity and fitting ability. However, MLP also suffers from the overfitting problem and requires a suitable regularization method to constrain the learning process of the neural network. Bhardwaj et al. [77] developed a semi-supervised learning method based on an adaptive MLP algorithm for long-range portable LIBS systems. The method is lightweight, low computational, and power cost, and the model training and prediction can be deployed on commercial cell phones with accuracy of 89.3% for quantitative analysis. Effects from experimental parameters and physical and mechanical properties of samples are studied [78] by designing the network inputs, including the laser single-pulse energy, the iCCD time delay, ionization energy of the measured element, the central wavelength of the corresponding spectral line, the concentration of individual elements, the reflectivity of the sample, the latent heat of vaporization, the specific heat, and the boiling point, in the MLP network model, which is used to obtain the optimal SNR of the LIBS spectra. The results show that the experimental parameters can be optimized efficiently using the MLP method, so it is possible to select the experimental parameter corresponding to the best SNR value as the optimization parameter for LIBS measurements and improves the performance of LIBS measurements. Both BPNN and MLP are shallow neural network algorithms, and they are highly adaptive, flexible, and predictive accurate. When choosing a specific algorithm, it is necessary to consider the data volume, sample characteristics, prediction accuracy, and other factors and choose a suitable neural network structure according to the actual situation.

4.3 Convolutional neural network (CNN)

CNN is a deep neural network method commonly used in quantitative LIBS analysis. Compared with the BPNN and MLP networks, the modules of convolution and pooling in the CNN framework achieve better performance in processing nonlinear LIBS spectral data. The basic principle of CNN in quantitative LIBS is to extract the internal correlation of the neighboring data points through the convolution operation to learn the deeper features in the LIBS spectrum.

Due to the rapid development of CNN models, CNN methods have been a hot research topic in quantitative LIBS research in recent years. Yang [79] and Li [80] et al. studied LIBS spectra collected during the preflight test of the Mars mission based on the Mars Surface Composition Detector (MarSCoDe) on Tianwen 1 Mars Rover, achieved high-precision quantitative LIBS by using a CNN model with five convolutional layers and two pooling layers. Castorena et al. [81] proposed a CNN network dedicated to spectral processing, which can simultaneously realize the preprocessing and quantitative measurement of LIBS spectral signals, and the experimental results demonstrated that this method was significantly better than the existing method used by the U.S. Mars rover Curiosity. Davari et al. [82] fused the Lasso model with a one-dimensional CNN method to analyze LIBS spectra of single-crystal silicon and measured the oxygen-related impurities in the samples with LOD as low as 0–16 ppm. Cui et al. [83] proposed a migration-learning multitask regularized CNN model for quantitative analysis of coal samples on a dataset with small sample size. Compared with PLS, SVM regression, and ordinary CNN networks, the RMSE of this method was reduced by 19.9%, 5.9%, and 7.7%, respectively. Choi et al. [84] used LIBS to monitor laser cleaning of painted stainless steel and proposed a deep learning method based on CNN for quantitatively analyzing whether the base elemental material appeared in LIBS spectra and achieved satisfied results in terms of real time and accuracy. Eynde et al. [85] used the CNN-architected GHOSTNET network for trace elements such as Fe, Cu, Mn, Mg, and Zn, which are found in very low levels in fertilizers, and compared it with the BPNN method. The results show that the average

RMSE of the deep learning method does not exceed up to 0.01%. In addition, the method is lightweight and real time with processing time only 10 ms, which plays an important role for the practical application of LIBS technology in metal sorting and other fields.

In conclusion, CNN can mine the intrinsic features and distribution of data and is a more accurate quantitative LIBS method. In practical applications, the appropriate network structure and training strategy can be selected according to the data characteristics and the constraints of computational resources to improve the accuracy and robustness of quantitative analysis.

In addition to the above commonly used neural networks, there are other deep learning methods that have been applied to LIBS quantitative analysis. For example, Wei et al. [86] used the wavelet neural network (WNN) to quantitatively analyze the main components in coal ash, and the results showed that WNN still has high accuracy in online detection. Rezaei et al. [87] used recurrent neural network (RNN) to compare the accuracy of quantitative analysis of aluminum alloy with other methods, such as SVM regression and MLP, and the results showed that the RNN had good performance for most of the element concentrations with the highest efficiency in quantitative measurements.

5. Discussion

Table 2 lists the comparative studies conducted on different machine learning methods on the same sample in recent years. We can see that the PLS algorithm is frequently used in the quantitative LIBS due to its simple implementation and good fitting. In addition, in most cases, the LSSVM method, which is based on an improved SVM, has greater fitness and smaller error.

It is worth noting that few studies have been conducted to compare and analyze the quantitative analysis of conventional machine learning methods and ANN methods due to the lack of benchmark quantitative LIBS datasets. It is well-known that for machine learning, especially deep learning methods, datasets play an important role in experimental studies such as model training, and the number and quality of datasets can significantly affect the accuracy of quantitative results. The current targets for quantitative LIBS based on machine learning include geological samples such as soil and rocks, alloy industrial materials, biomedical samples, food materials, and many other target samples with different properties, elemental contents, shapes, and states. These researches cover a wide range of fields, most of the datasets are not open source nor public available, and many research groups have private datasets, so it is difficult to compare the performance of different machine learning methods fairly in analogy to the fields of image processing and machine vision, which have many open-source public datasets. Kepes et al. [95] have previously proposed and created a benchmark dataset of LIBS for soil for classification tasks and launched a LIBS classification competition at the EMSLIBS 2019 conference [96]. The dataset contains tens of thousands of LIBS spectral data from a total of 138 soil samples of 12 different species. Unfortunately, however, there is no publicly recognized benchmark dataset for quantitative LIBS analysis to compare the performance of different machine learning algorithms.

With the rapid development of artificial intelligence field, LIBS quantitative analysis is facing opportunities. In the following, we throw light on the development of AI+LIBS from the aspects of mechanism, data, and method to provide viewpoint support for the practicality of high-precision quantitative LIBS.

Samples	Target elements/compound	Method	RMSEP	R^2
Iron ore [88]	Fe, SiO_2, Al_2O_3, CaO, MgO	PLS	1.218 wt%	0.9710
		LSSVM	0.624 wt%	0.9933
Iron Ore [89]	CaO, MgO, SiO_2, Al_2O_3	PLS	7.8018 wt%	0.7166
		LSSVM	6.5724 wt%	0.8047
Coal [90]	C, Si, Al, Fe, Ca, Na, K	PLS	1.68 mJ/kg	0.9300
		SVM	1.08 J/kg	0.9700
Pork [71]	Cr	PLS	4.8362 wt%	0.9696
		SVM	0.6961 wt%	0.9973
		MLP	15.2996 wt%	0.9136
Olivine [91]	Mg_2SiO_4, Fe_2SiO_4	MLP	29.03	0.9010
		BPNN	28.64	0.9110
Chemical fertilizer [92]	N, P, K	SVM	0.0752	0.9656
		LSSVM	0.0242	0.9960
Soil [93]	Cd	PLS	0.051	0.9463
		LSSVM	0.034	0.9730
Cement [94]	Ca, Si, Al	PLS	1.3517%	0.8260
		SVM	0.5831%	0.9900

Table 2.
Summary of a comparative study of different machine learning methods on the same sample.

5.1 Spectral data acquisition based on LIBS mechanism

The state of LIBS plasma is crucial for quantitative analysis, and to make LIBS data measurement more accurate, the spectral signals need to be collected at the appropriate time, orientation, environment, and other experimental parameters in the experiment. However, the kinetic mechanism of the emission spectra during the time-dependent expansion of LIBS plasma is complicated, and most of the previous works have not carefully studied the selection of experimental parameters, resulting in poor accuracy of the raw LIBS spectral data. Some recent works used machine learning methods for optimization [78] but did not give a suitable spatial and temporal range for sampling LIBS spectral measurements from the mechanism. To address this, in recent years, the authors' group has conducted several studies on the evolution of LIBS plasma dynamics. Firstly, we diagnosed the silicon plasma evolution process and parameters such as electron temperature and density by a temporal–spatial-spectral resolved observation system and found that it is difficult for the plasma to reach the local thermodynamic equilibrium (LTE) within the first 200 ns after laser ablation, and we derived a sufficient and necessary criterion for LTE from an universal electron energy distribution function, which is more effective than the McWhirter criterion [97], which provides theoretical and technical methods for the measurement of LIBS plasma parameter evolution with time and space [25]. Subsequently, in order to enrich the parameter database required for quantitative analysis of LIBS plasma, a method of expanding the database using cross-correction was proposed, which promotes the practicability of quantitative analysis of LIBS plasma in the non-LTE [26]. In order to analyze the mechanism of wavelength-dependent LIBS, the authors [17] fused a hydrodynamics-based adiabatic expansion simulation model and spatial–temporal spectrally resolved experimental observations to reveal the mechanism by which plasmas generated by long-wavelength lasers are more likely to satisfy LTE. The experimental results show that the LIBS plasma has stronger inverse bremsstrahlung absorption for 1064 nm nanosecond pulsed lasers compared to 532 nm lasers, and the higher electron temperature leads to a more rapid energy transfer within the plasma and thus is easier to reach the LTE state. Recently, the author's team has established a temporal–spatial resolved thermodynamic state evolution model of LIBS plasma and proposed the LTE criterion based on the spectral line fitting method, which realizes a more accurate identification of the time window and spatial location of the LIBS plasma in LTE. The above mechanism studies provide a theoretical basis for the accurate acquisition of LIBS spectral data. For example, Ke et al. [98] used this criterion to optimize the acquisition time of LIBS signals under different background air pressures, which significantly increased the accuracy of quantitative LIBS under different air pressure conditions.

5.2 Data issues

As mentioned in Section 4.3, sufficient and high-quality data can train high-performing machine learning models. However, the lack of open-source large-scale datasets is still a difficulty for community of LIBS. On the one hand, we should continue to call for the establishment of publicly available benchmark datasets in the field of LIBS research, and on the other hand, we can find ways to overcome the bottleneck from the interdisciplinary aspect. Few-shot learning is not only a challenge in the field of LIBS quantitative analysis but also a frontier and hotspot in the current development of new generation artificial intelligence. Methods, such as data augmentation, transfer learning, and deep learning with the fusion of knowledge-guided and data-driven, have been proposed in recent research in the field of AI to solve the small-sample problem in practical applications. These studies can provide ideas and inspiration for LIBS quantitative analysis. For example, in a recently published work, Cui et al. [83] proposed a transfer learning multitask regularized CNN model to achieve quantitative analysis of coal samples on a small number of target samples. In the future, we can introduce constraints such as multitask regularization and make full use of the physical knowledge embedded in the spectral data as a prior information, which is expected to achieve high-precision LIBS analysis in the case of small samples.

5.3 Interpretability of deep learning

Although deep learning has demonstrated excellent performance for nonlinear and complex problems in many fields, including LIBS quantitative analysis, its "black box" characteristics make safe, trustworthy, and interpretable deep learning methods a key issue in the development of the new generation of AI. Since there is a theoretical mapping relationship between LIBS spectra and elemental concentrations based on physicochemical mechanisms, although the coupling of matrix effects, self-absorption, plasma evolution, and other complex factors leads to the nonlinearization of this mapping relationship, the deep learning methods in quantitative LIBS still have certain interpretability. For example, it has been shown [80] that the quantization ability of PLS can be greatly improved by using baseline removal as a spectral data preprocessing step, while the quantization ability of BPNN is almost unchanged, and the quantization ability of CNN is even slightly decreased, and experiments have shown that the performance of full-spectrum-based CNN quantization in [99] is better than the prediction of CNN based on selected spectral regions. These results indicate that although there is a large amount of noise signal in the baseline, it still contains useful information making the output of the deep learning model more accurate. The underlying reason is that although the LIBS line spectrum is the main feature of raw data, the continuous spectrum comes from physical processes, such as bremsstrahlung, which implies information, such as the plasma temperature, so the CNN model using the full spectral information obtains more. In the future, the knowledge of physics and chemometrics applied to spectroscopic analysis can be combined with neural networks to innovate network architecture, training process, loss function, and the use of methods, such as supervised and reinforcement learning independently or jointly. Utilizing knowledge-inspired deep learning methods can not only solve the small sample but also enhance the interpretability of the deep learning methods, serving as an example of the interpretability of next-generation AI technologies while promoting further improvement of the accuracy of quantitative LIBS.

6. Conclusion

This chapter provides an overview of the current bottlenecks in LIBS quantitative analysis, outlines the common-used data preprocessing methods required for machine learning (ML), and reviews the research status and progress of conventional ML methods such as PLS, SVM, Lasso, and ANN-based quantitative analysis methods. Finally, in response to challenges and opportunities in the development of AI and LIBS, suggestions are proposed from the perspectives of mechanism, dataset problems, and interpretability of deep learning. Through this review, we aim to provide support for interdisciplinary research of AI and LIBS.

Acknowledgements

The authors H Liu and WQ Yang would like to acknowledgethe funding support of the pre-research project on Civil Aerospace Technologies funded by China National Space Administration (Grant No. D010105).

DOI: http://dx.doi.org/10.5772/intechopen.1004414

Author details

Hao Liu[1,2], Kai Han[1,2*], Weiqiang Yang[1,2*] and Minsun Chen[1,2]

1 College of Advanced Interdisciplinary Studies, National University of Defense Technology, Changsha, China

2 Nanhu Laser Laboratory, National University of Defense Technology, Changsha, China

*Address all correspondence to: hankai0071@nudt.edu.cn
and yangweiqiang_001@126.com

References

[1] Al-Najjar OA, Wudil YS, Ahmad UF, et al. Applications of laser induced breakdown spectroscopy in geotechnical engineering: A critical review of recent developments, perspectives and challenges. Applied Spectroscopy Reviews. 2022;**58**(10):687-723

[2] Senesi G, Harmon R, Hark R. Field-portable and handheld laser-induced breakdown spectroscopy: Historical review, current status and future prospects. Spectrochimica Acta Part B Atomic Spectroscopy. 2020;**175**:106013

[3] Li L-N, Liu X-F, Yang F, et al. A review of artificial neural network based chemometrics applied in laser-induced breakdown spectroscopy analysis. Spectrochimica Acta Part B: Atomic Spectroscopy. 2021;**180**:106183

[4] Zhang D, Zhang H, Zhao Y, et al. A brief review of new data analysis methods of laser-induced breakdown spectroscopy: Machine learning. Applied Spectroscopy Reviews. 2020;**57**:1-23

[5] Imran M, Hu Z, Ding F, et al. Diagnostic study of impurity deposition in fusion device by calibration-free laser-induced breakdown spectroscopy. Spectrochimica Acta Part B: Atomic Spectroscopy. 2022;**198**:106568

[6] van der Meiden HJ, Almaviva S, Butikova J, et al. Monitoring of tritium and impurities in the first wall of fusion devices using a LIBS based diagnostic. Nuclear Fusion. 2021;**61**(12):125001

[7] Li C, Feng C-L, Oderji HY, et al. Review of LIBS application in nuclear fusion technology. Frontiers of Physics. 2016;**11**(6):114214

[8] Pedarnig JD, Trautner S, Grünberger S, et al. Review of element analysis of industrial materials by In-line laser—Induced breakdown spectroscopy (LIBS). Applied Sciences. 2021;**11**(19):9274

[9] Legnaioli S, Campanella B, Poggialini F, et al. Industrial applications of laser-induced breakdown spectroscopy: A review. Analytical Methods. 2020;**12**(8):1014-1029

[10] Képeš E, Vrábel J, Siozos P, et al. Quantification of alloying elements in steel targets: The LIBS 2022 regression contest. Spectrochimica Acta Part B: Atomic Spectroscopy. 2023;**206**:106710

[11] Zhang D, Zhang H, Zhao Y, et al. A brief review of new data analysis methods of laser-induced breakdown spectroscopy: Machine learning. Applied Spectroscopy Reviews. 2022;**57**(2):89-111

[12] Galbács G, Kéri A, Kohut A, et al. Nanoparticles in analytical laser and plasma spectroscopy – A review of recent developments in methodology and applications. Journal of Analytical Atomic Spectrometry. 2021;**36**(9):1826-1872

[13] Dell'Aglio M, Alrifai R, De Giacomo A. Nanoparticle enhanced laser induced breakdown spectroscopy (NELIBS), a first review. Spectrochimica Acta Part B: Atomic Spectroscopy. 2018;**148**:105-112

[14] Matsumoto A, Sakka T. A review of underwater laser-induced breakdown spectroscopy of submerged solids. Analytical Sciences. 2021;**37**(8):1061-1072

[15] Ji H, Ding Y, Zhang L, et al. Review of aerosol analysis by laser-induced breakdown spectroscopy.

Applied Spectroscopy Reviews. 2021;**56**(3):193-220

[16] Kautz EJ, Weerakkody EN, Finko MS, et al. Optical spectroscopy and modeling of uranium gas-phase oxidation: Progress and perspectives. Spectrochimica Acta Part B: Atomic Spectroscopy. 2021;**185**:106283

[17] Liu H, Ashfold MNR, Meehan DN, et al. Wavelength-dependent variations of the electron characteristics in laser-induced plasmas: A combined hydrodynamic and adiabatic expansion modelling and time-gated, optical emission imaging study. Journal of Applied Physics. 2019;**125**(8):083304

[18] Liu Z, Zhao G, Guo C, et al. Spatially and temporally resolved evaluation of local thermodynamic equilibrium for laser-induced plasma in a high vacuum. Journal of Analytical Atomic Spectrometry. 2021;**36**(11):2362-2369

[19] Chen T, Zhang T, Li H. Applications of laser-induced breakdown spectroscopy (LIBS) combined with machine learning in geochemical and environmental resources exploration. TrAC Trends in Analytical Chemistry. 2020;**133**:116113

[20] Huang Y, Harilal SS, Bais A, et al. Progress toward machine learning methodologies for laser-induced breakdown spectroscopy with an emphasis on soil analysis. IEEE Transactions on Plasma Science. 2022;**51**(7):1729-1749

[21] Harilal SS, Brumfield BE, LaHaye NL, et al. Optical spectroscopy of laser-produced plasmas for standoff isotopic analysis. Applied Physics Reviews. 2018;**5**(2):021301

[22] Kautz EJ, Yeak J, Bernacki BE, et al. The role of ambient gas confinement, plasma chemistry, and focusing conditions on emission features of femtosecond laser-produced plasmas. Journal of Analytical Atomic Spectrometry. 2020;**35**(8):1574-1586

[23] Kautz EJ, Senor DJ, Harilal SS. The interplay between laser focusing conditions, expansion dynamics, ablation mechanisms, and emission intensity in ultrafast laser-produced plasmas. Journal of Applied Physics. 2021;**130**(20):204901

[24] Liu Z, Guo C, Chen L, et al. Thermodynamic equilibrium state analysis of silicon plasma induced by picosecond laser. In: Proceedings of the 7th Asia Pacific Conference on Optics Maufacture, SPIE, 12166. 2022

[25] Liu H, Truscott BS, Ashfold MNR. Position- and time-resolved stark broadening diagnostics of a non-thermal laser-induced plasma. Plasma Sources Science and Technology. 2016;**25**(1):015006

[26] Liu H, Truscott BS, Ashfold MNR. Determination of stark parameters by cross-calibration in a multi-element laser-induced plasma. Scientific Reports. 2016;**6**(1):25609

[27] Takahashi T, Thornton B. Quantitative methods for compensation of matrix effects and self-absorption in laser induced breakdown spectroscopy signals of solids. Spectrochimica Acta Part B: Atomic Spectroscopy. 2017;**138**:31-42

[28] Fu X, Li G, Dong D. Improving the detection sensitivity for laser-induced breakdown spectroscopy: A review. Frontiers in Physics. 2020;**8**. Article no.: 68

[29] Ahmed R, Jabbar A, Akhtar M, et al. Amelioration in the detection of chlorine

using electric field assisted LIBS. Plasma Chemistry and Plasma Processing. 2020;**40**(4):809-818

[30] Wu D, Sun L, Hai R, et al. Influence of transverse magnetic field on plume dynamics and optical emission of nanosecond laser produced tungsten plasma in vacuum. Spectrochimica Acta Part B: Atomic Spectroscopy. 2020;**169**:105882

[31] Wubetu GA, Fiedorowicz H, Costello JT, et al. Time resolved anisotropic emission from an aluminium laser produced plasma. Physics of Plasmas. 2017;**24**(1):013105

[32] Tang Z, Liu K, Hao Z, et al. The validity of nanoparticle enhanced molecular laser-induced breakdown spectroscopy. Journal of Analytical Atomic Spectrometry. 2021;**36**(5):1034-1040

[33] Guezenoc J, Gallet-Budynek A, Bousquet B. Critical review and advices on spectral-based normalization methods for LIBS quantitative analysis. Spectrochimica Acta Part B: Atomic Spectroscopy. 2019;**160**:105688

[34] Dell'Aglio M, Gaudiuso R, Senesi GS, et al. Monitoring of Cr, Cu, Pb, V and Zn in polluted soils by laser induced breakdown spectroscopy (LIBS). Journal of Environmental Monitoring. 2011;**13**(5):1422-1426

[35] Fabre C, Cousin A, Wiens RC, et al. In situ calibration using univariate analyses based on the onboard ChemCam targets: First prediction of Martian rock and soil compositions. Spectrochimica Acta Part B: Atomic Spectroscopy. 2014;**99**:34-51

[36] Syvilay D, Wilkie-Chancellier N, Trichereau B, et al. Evaluation of the standard normal variate method for laser-induced breakdown spectroscopy data treatment applied to the discrimination of painting layers. Spectrochimica Acta Part B: Atomic Spectroscopy. 2015;**114**:38-45

[37] Wang R, Ma X, Zhang T, Liu Z, Huo L. Study on the data processing method applied to improve spectral stability of laser induced breakdown spectroscopy in soil analysis. Applied Optics and Photonics China. 2019

[38] Thomas et al. Characterization of hydrogen in basaltic materials with Laser-Induced Breakdown Spectroscopy (LIBS) for application to MSL ChemCam data. Journal of Geophysical Research: Planets. 2018;**123**. DOI: 10.1029/2017JE005467

[39] Andrade et al. Calibration strategies for determination of the content in discarded liquid crystal displays (LCD) from mobile phones using laser-induced breakdown spectroscopy (LIBS). Analytica Chimica Acta. 2019. DOI: 10.1016/j.aca.2019.02.038

[40] Payre et al. Alkali trace elements in Gale crater, Mars, with ChemCam: Calibration update and geological implications. Journal of Geophysical Research: Planets. 2017;**122**:431-684

[41] Takahashi et al. Partial least squares regression calculation for quantitative analysis of metals submerged in water measured using laser-induced breakdown spectroscopy. Applied Optics. 2018;**57**(20)

[42] Lazic V, Trujillo-Vazquez A, Sobral H, et al. Corrections for variable plasma parameters in laser induced breakdown spectroscopy: Application on archeological samples. Spectrochimica Acta Part B: Atomic Spectroscopy. 2016;**122**:103-113

[43] Feng J, Wang Z, Li Z, et al. Study to reduce laser-induced breakdown

spectroscopy measurement uncertainty using plasma characteristic parameters. Spectrochimica Acta Part B: Atomic Spectroscopy. 2010;**65**(7):549-556

[44] Li L, Wang Z, Yuan T, et al. A simplified spectrum standardization method for laser-induced breakdown spectroscopy measurements. Journal of Analytical Atomic Spectrometry. 2011;**26**(11):2274-2280

[45] Sarkar A, Karki V, Aggarwal SK, et al. Evaluation of the prediction precision capability of partial least squares regression approach for analysis of high alloy steel by laser induced breakdown spectroscopy. Spectrochimica Acta Part B: Atomic Spectroscopy. 2015;**108**:8-14

[46] Castro JP, Pereira-Filho ER. Twelve different types of data normalization for the proposition of classification, univariate and multivariate regression models for the direct analyses of alloys by laser-induced breakdown spectroscopy (LIBS). Journal of Analytical Atomic Spectrometry. 2016;**31**(10):2005-2014

[47] dos Santos AA, Barsanelli PL, Pereira FMV, et al. Calibration strategies for the direct determination of Ca, K, and Mg in commercial samples of powdered milk and solid dietary supplements using laser-induced breakdown spectroscopy (LIBS). Food Research International. 2017;**94**:72-78

[48] Zhang P, Sun L, Yu H, et al. An image auxiliary method for the quantitative analysis of laser-induced breakdown spectroscopy. Analytical Chemistry. 2018;**90**:4686-4694

[49] Davies CM, Telle HH, Montgomery DJ, et al. Quantitative analysis using remote laser-induced breakdown spectroscopy (LIBS). Spectrochimica Acta Part B: Atomic Spectroscopy. 1995;**50**(9):1059-1075

[50] Bredice F, Borges FO, Sobral H, et al. Evaluation of self-absorption of manganese emission lines in laser induced breakdown spectroscopy measurements. Spectrochimica Acta Part B: Atomic Spectroscopy. 2006;**61**(12):1294-1303

[51] Gornushkin IB, Anzano JM, King LA, et al. Curve of growth methodology applied to laser-induced plasma emission spectroscopy. Spectrochimica Acta Part B: Atomic Spectroscopy. 1999;**54**(3):491-503

[52] Kadachi AN, Al-Eshaikh MA, Ahmad K. Self-absorption correction: An effective approach for precise quantitative analysis with laser induced breakdown spectroscopy. Laser Physics. 2018;**28**(9):095701

[53] Lazic V, Barbini R, Colao F, et al. Self-absorption model in quantitative laser induced breakdown spectroscopy measurements on soils and sediments. Spectrochimica Acta Part B: Atomic Spectroscopy. 2001;**56**(6):807-820

[54] Anderson RB, Clegg SM, Frydenvang J, et al. Improved accuracy in quantitative laser-induced breakdown spectroscopy using sub-models. Spectrochimica Acta Part B: Atomic Spectroscopy. 2017;**129**:49-57

[55] Yaroshchyk P, Death DL, Spencer SJ. Comparison of principal components regression, partial least squares regression, multi-block partial least squares regression, and serial partial least squares regression algorithms for the analysis of Fe in iron ore using LIBS. Journal of Analytical Atomic Spectrometry. 2012;**27**(1):92-98

[56] Yuan T, Wang Z, Li Z, et al. A partial least squares and wavelet-transform hybrid model to analyze carbon content in coal using laser-induced breakdown

spectroscopy. Analytica Chimica Acta. 2014;**807**:29-35

[57] Li A, Guo S, Wazir N, et al. Accuracy enhancement of laser induced breakdown spectra using permittivity and size optimized plasma confinement rings. Optics Express. 2017;**25**(22):27559-27569

[58] Doucet FR, Lithgow G, Kosierb R, et al. Determination of isotope ratios using laser-induced breakdown spectroscopy in ambient air at atmospheric pressure for nuclear forensics. Journal of Analytical Atomic Spectrometry. 2011;**26**(3):536-541

[59] Takahashi T, Thornton B, Sato T, et al. Temperature based segmentation for spectral data of laser-induced plasmas for quantitative compositional analysis of brass alloys submerged in water. Spectrochimica Acta Part B: Atomic Spectroscopy. 2016;**124**:87-93

[60] Wiens RC, Maurice S, Barraclough B, et al. The ChemCam instrument suite on the Mars science laboratory (MSL) rover: Body unit and combined system tests. Space Science Reviews. 2012;**170**(1):167-227

[61] Meslin PY, Gasnault O, Forni O, et al. Soil diversity and hydration as observed by ChemCam at Gale crater, Mars. Science. 2013;**341**(6153):1238670

[62] Tucker J, Dyar M, Schaefer M, et al. Optimization of laser-induced breakdown spectroscopy for rapid geochemical analysis. Chemical Geology – CHEM GEOL. 2010;**277**:137-148

[63] Clegg SM, Sklute E, Dyar MD, et al. Multivariate analysis of remote laser-induced breakdown spectroscopy spectra using partial least squares, principal component analysis, and related techniques. Spectrochimica Acta Part B: Atomic Spectroscopy. 2009;**64**(1):79-88

[64] Anderson R, Iii J, Wiens R, et al. Clustering and training set selection methods for improving the accuracy of quantitative laser induced breakdown spectroscopy. Spectrochimica Acta Part B Atomic Spectroscopy. 2012;**70**:24-32

[65] Wiens RC, Maurice S, Lasue J, et al. Pre-flight calibration and initial data processing for the ChemCam laser-induced breakdown spectroscopy instrument on the Mars science laboratory rover. Spectrochimica Acta Part B: Atomic Spectroscopy. 2013;**82**:1-27

[66] Dyar MD, Carmosino ML, Breves EA, et al. Comparison of partial least squares and lasso regression techniques as applied to laser-induced breakdown spectroscopy of geological samples. Spectrochimica Acta Part B: Atomic Spectroscopy. 2012;**70**:51-67

[67] Boucher TF, Ozanne MV, Carmosino ML, et al. A study of machine learning regression methods for major elemental analysis of rocks using laser-induced breakdown spectroscopy. Spectrochimica Acta Part B: Atomic Spectroscopy. 2015;**107**:1-10

[68] Yao S, Xu J, Bai K, et al. Improved measurement performance of inorganic elements in coal by laser-induced breakdown spectroscopy coupled with internal standardization*. Plasma Science and Technology. 2015;**17**(11):938

[69] Wang X, Zhang L, Fan J, et al. Parameters optimization of laser-induced breakdown spectroscopy experimental setup for the case with beam expander*. Plasma Science and Technology. 2015;**17**(11):914

[70] Shi Q, Niu G, Lin Q, et al. Quantitative analysis of sedimentary

rocks using laser-induced breakdown spectroscopy: Comparison of support vector regression and partial least squares regression chemometric methods. Journal of Analytical Atomic Spectrometry. 2015;**30**(12):2384-2393

[71] Chen T, Zhang L, Huang L, et al. Quantitative analysis of chromium in pork by PSO-SVM chemometrics based on laser induced breakdown spectroscopy. Journal of Analytical Atomic Spectrometry. 2019;**34**(5):884-890

[72] Dai Y, Song C, Gao X, et al. Quantitative determination of Al–Cu–Mg–Fe–Ni aluminum alloy using laser-induced breakdown spectroscopy combined with LASSO–LSSVM regression. Journal of Analytical Atomic Spectrometry. 2021;**36**(8):1634-1642

[73] Zhang X, Li N, Yan C, et al. Four-metal-element quantitative analysis and pollution source discrimination in atmospheric sedimentation by laser-induced breakdown spectroscopy (LIBS) coupled with machine learning. Journal of Analytical Atomic Spectrometry. 2020;**35**(2):403-413

[74] Zhang Y, Sun C, Zengqi Y, et al. Correlation-based carbon determination in steel without explicitly involving carbon-related emission lines in a LIBS spectrum. Optics Express. 2020;**28**:32019

[75] Wang R, Ma X. Study on LIBS standard method via key parameter monitoring and backpropagation neural network. Chem. 2022;**10**(8):312

[76] Kohonen T. An introduction to neural computing. Neural Networks. 1988;**1**(1):3-16

[77] Bhardwaj K, Gokhale M. Semi-supervised on-device neural network adaptation for remote and portable laser-induced breakdown spectroscopy. ArXiv. 2021. ArXiv: abs/2104.03439

[78] Prochazka D, Pořízka P, Hruška J, et al. Machine learning in laser-induced breakdown spectroscopy as a novel approach towards experimental parameter optimization. Journal of Analytical Atomic Spectrometry. 2022;**37**(3):603-612

[79] Yang F, Li L-N, Xu W-M, et al. Laser-induced breakdown spectroscopy combined with a convolutional neural network: A promising methodology for geochemical sample identification in Tianwen-1 Mars mission. Spectrochimica Acta Part B: Atomic Spectroscopy. 2022;**192**:106417

[80] Li L-N, Liu X-F, Xu W-M, et al. A laser-induced breakdown spectroscopy multi-component quantitative analytical method based on a deep convolutional neural network. Spectrochimica Acta Part B: Atomic Spectroscopy. 2020;**169**:105850

[81] Castorena J, Oyen D, Ollila A, et al. Deep spectral CNN for laser induced breakdown spectroscopy. Spectrochimica Acta Part B: Atomic Spectroscopy. 2021;**178**:106125

[82] Davari SA, Mukherjee D. Deep learning models for data-driven laser induced breakdown spectroscopy (LIBS) analysis of interstitial oxygen impurities in Czochralski-Si crystals. Applied Spectroscopy. 2022;**76**(6):667-677

[83] Cui J, Song W, Hou Z, et al. A transferred multitask regularization convolutional neural network (TrMR-CNN) for laser-induced breakdown spectroscopy quantitative analysis. Journal of Analytical Atomic Spectrometry. 2022;**37**(10):2059-2068

[84] Choi S, Park C. Convolution neural network with laser-induced breakdown spectroscopy as a monitoring tool for laser cleaning process. Sensors. 2023;**23**(1):83

[85] Van den Eynde S, Díaz-Romero DJ, Zaplana I, et al. Deep learning regression for quantitative LIBS analysis. Spectrochimica Acta Part B: Atomic Spectroscopy. 2023;**202**:106634

[86] Wei J, Dong J, Zhang T, et al. Quantitative analysis of the major components of coal ash using laser induced breakdown spectroscopy coupled with a wavelet neural network (WNN). Analytical Methods. 2016;**8**(7):1674-1680

[87] Rezaei F, Khalilian P, Rezaei M, Karimi P, Ashrafkhani B. A comparison between Recurrent Neural Networks and classical machine learning approaches in laser induced breakdown spectroscopy. ArXiv. 2023. ArXiv: abs/2304.08500

[88] Guo YM, Guo LB, Hao ZQ, et al. Accuracy improvement of iron ore analysis using laser-induced breakdown spectroscopy with a hybrid sparse partial least squares and least-squares support vector machine model. Journal of Analytical Atomic Spectrometry. 2018;**33**(8):1330-1335

[89] Wang P, Li N, Yan C, et al. Rapid quantitative analysis of the acidity of iron ore by the laser-induced breakdown spectroscopy (LIBS) technique coupled with variable importance measures-random forests (VIM-RF). Analytical Methods. 2019;**11**(27):3419-3428

[90] Li X, Yang Y, Li G, et al. Accuracy improvement of quantitative analysis of calorific value of coal by combining support vector machine and partial least square methods in laser-induced breakdown spectroscopy. Plasma Science and Technology. 2020;**22**(7):074014

[91] Ru-jun Y, Xiong W, Qiang H, et al. Research on olivine component analysis using LIBS combined with Back-propagation algorithm. Spectroscopy and Spectral Analysis. 2019;**39**(12):7

[92] Sha W, Li J, Xiao W, et al. Quantitative analysis of elements in fertilizer using laser-induced breakdown spectroscopy coupled with support vector regression model. Sensors. 2019;**19**(15):3277

[93] Liu X, Liu F, Huang W, et al. Quantitative determination of Cd in soil using laser-induced breakdown spectroscopy in air and Ar conditions. Molecules. 2018;**23**:2492

[94] Zhi-wei G, Lan-xiang S, Peng Z, et al. On-line component analysis of cement powder using LIBS technology. Spectroscopy and Spectral Analysis. 2019;**39**(01):278-285. DOI: 10.3964/j.issn.1000-0593(2019)01-0278-08

[95] Képeš E, Vrábel J, Střítežská S, et al. Benchmark classification dataset for laser-induced breakdown spectroscopy. Scientific Data. 2020;**7**(1):53

[96] Vrábel J, Képeš E, Duponchel L, et al. Classification of challenging laser-induced breakdown spectroscopy soil sample data – EMSLIBS contest. Spectrochimica Acta Part B: Atomic Spectroscopy. 2020;**169**:105872

[97] Griem HR. Principles of Plasma Spectroscopy. Cambridge: Cambridge University Press; 1997

[98] Ke W, Wang X, Chen M, et al. Influence of ambient pressure on spatial–temporal evolution of local thermodynamic equilibrium for laser-induced plasma. Journal of Analytical Atomic Spectrometry. 2023;**38**(1):212-220

[99] He Y, Zhao Y, Zhang C, et al. Discrimination of grape seeds using laser-induced breakdown spectroscopy in combination with region selection and supervised classification methods. Food. 2020;**9**(2):199